A James Martin BOOK

THE WIRED SOCIETY

Other Books by James Martin

Principles of Data-Base Management
Computer Data-Base Organization, 2nd Ed.
Security, Accuracy, and Privacy in Computer Systems
Design of Man-Computer Dialogues
Introduction to Teleprocessing
Teleprocessing Network Organization
Systems Analysis for Data Transmission
Telecommunications and the Computer, 2nd Ed.
Future Developments in Telecommunications, 2nd Ed.
The Computerized Society
Design of Real-Time Computer Systems
Programming Real-Time Computer Systems

James Martin

THE WIRED SOCIETY

PRENTICE-HALL, INC.

Englewood Cliffs, New Jersey

The Wired Society by James Martin

Printed in the United States of America
Prentice-Hall International, Inc., London
Prentice-Hall of Australia, Pty. Ltd., Sydney
Prentice-Hall of Canada, Ltd., Toronto
Prentice-Hall of India Private Ltd., New Delhi
Prentice-Hall of Japan, Inc., Tokyo
Prentice-Hall of Southeast Asia Pte. Ltd., Singapore
Whitehall Books Limited, Wellington, New Zealand
10 9 8 7 6 5 4 3 2 1

Library of Congress Cataloging in Publication Data

Martin, James,
The wired society.
Includes bibliographical references and index.
1. Telecommunication. 2. Computers. 3. Electronic data processing. I. Title.
TK5101.M327 384 77-18087
ISBN 0-13-961441-9

To Charity

CONTENTS

PREFACE

In the past, communications networks have been built for telephony, telegraphy, and broadcasting. Now the technology of communications is changing in ways which will have impact on the entire fabric of society in both developed and developing nations. In the U.S.A. the technology revolution coincides with a change in the political and legal structure of the telecommunications industry; the combination is explosive.

Some countries will take advantage of the new technology; some will not. Some businessmen will make fortunes. Some companies will be bankrupted.

Faced with the dilemmas of our age it is imperative that we employ benevolent technologies to tackle our problems and build a better world. The decisions to do so or not will be made partly by politicians and partly by the marketplace. Young persons especially will be affected by it. Only through widespread understanding of the potentials can man make a choice about the future.

The book is written without technical language. For readers who want information about the technology, the author wrote two other books at the same time: *Future Developments in Telecommunications* (a completely rewritten second edition) and *Communication Satellite Systems*.

There is a tide in the affairs of men,
Which, taken at flood, leads on to fortune,
On such a full sea are we now afloat,
And we must take the current when it serves,
Or lose our ventures.

WM. SHAKESPEARE
Julius Caesar

Chapter One

TECHNOLOGY AND THE ENVIRONMENT

As technology grows in power, its ability to either disrupt or heal increases. To heal we must move to new technologies, new social patterns, new types of consumer products, new ways of generating and spending wealth.

Technology is becoming an arch-villain in the minds of many.

Rivers are polluted. The fish in some areas are poisoned. More people die in road accidents than in earlier wars. The cities are foul with noise, traffic jams, and air pollution. Polar bears and whales may become extinct. We live in the shadow of nuclear weapons, drugs, and terrorism. The news media tell us that if aerosol sprays do not wreck the earth's ozone layer, the supersonic aircraft will.

Western society has become as dependent on mass energy consumption as an addict is on heroin. To remove the energy and make the addict go "cold turkey" would create convulsions more violent than civil war. Cities would be inoperable; economies would collapse; the mechanisms for distribution of food and goods would be paralysed. The tough and the violent would take matters into their own hands.

Yet oil is running out. There is less than thirty-five years' supply left at current consumption rates, and world consumption is rapidly growing. In the 1930s the United States found 275–300 barrels of crude oil for every foot of exploratory drilling. Now this figure has fallen to 15–25 barrels. To find alternative energy we are scarring the landscape to extract coal, building nuclear reactors of questionable safety, and planning breeder reactors that create cheap fuel for atomic bombs. Some countries plan to be major exporters of such breeder reactors in the 1980s.

Along with large-scale technology we have large-scale government. More and more citizens feel that they have no say in government. Their protests are not heard. Voting every few years seems to make no difference. The channels of communication are one-way: from the center to the citizens.

Some of our more extreme writers and environmentalists advocate an almost complete abandonment of technology. A back-to-nature movement has arisen, with vegetable-growing communes in California and primitive farmsteads in Vermont. However, the answer is not to abandon technology. The world population is now over 4 billion and will be at least 7 billion by the time our oil runs out. Without technology the earth cannot support more than about 2 billion people; to feed 4 billion requires highly advanced technology.

And we want to do more than feed the earth's people. Humanity's destiny is not mere subsistence. The future will bring not a return to primitive conditions of the past but a change to new forms of civilization. The new forms will have different consumption patterns, because along with oil many of the minerals needed to run today's industries are running out. We are frequently told that a civilization based on abundance is a thing of the past, and it is true that the era of cheap oil and cheap metals that have been the basis for industrial society is ending. There are, however, *new* forms of abundance which are rarely mentioned or understood. Microminiature electronic circuitry can provide unimaginable quantities of logic and computing power. We no longer argue about how many angels could be present on the head of a pin, but about how many transistors. Satellites and new glass fiber links can provide communications capacity that is truly abundant. Fusion and solar power will provide energy independent of exhaustible resources.

In fact, the research laboratories of the world have never been more vibrant with new technology than they are today. We are acquiring the capability to work new miracles, just as the steam engine and airplane were miracles to earlier ages. What faces us is not an end to Western abundance, because we are unlocking great new riches. Our problem is a problem of choice: Are we able to choose those technologies which will heal rather than spread the infections and avoid long-term harm?

As technology grows in power, its ability either to disrupt or to heal increases. We can destroy the planet more easily than we can heal the harm we have done so far. To heal, we have to move to new technologies, new social patterns, new types of consumer products, new ways of generating and spending wealth. Such changes will inevitably occur, whether they are brought by healing forethought or mindless destruction. The future will not be a repetition of the past.

In 1970 a group of students painted a word in twelve-foot letters on the road surface of First Avenue in New York

City so that it would stare up at the dignitaries high in the United Nations Building. The word was ECOLOGY. Until that time this word was little known; it is still absent from many dictionaries not recently revised. Until the late 1960s little thought was given to the tangled interrelations between industry and the environment, between consumption patterns and patterns of nature. Now the complexity of nature's web is better appreciated. Nature was once thought to be robust enough to withstand the worst that humankind could do. Now we know better. We have destroyed much and have reached a point where future destruction could be disastrous on a grand scale. The oceans *could* be dead. The ozone layer *could* be destroyed. We *could* create a radioactive nightmare. Or on a less cataclysmic level, there *could* be no butterflies; the New England maple trees *could* be wiped out as the elms were; eating shellfish *could* cause hepatitis; cancer incidence *could* increase tenfold.

The growth of the world's population to 7 billion will place an extreme strain on the earth's ecology. Affluence is growing in many developing nations, and the strain will be much greater if the burgeoning population demands consumer products like those of the West today. The earth could supply neither the energy nor the resources, nor could it absorb the pollution.

It is technology that has created this dilemma, and yet the only way out of the dilemma is more technology. To abandon technology or to stop further development would mean starvation on a scale the planet has never known before. Such abandonment will not happen and indeed is discussed seriously only in the very rich nations. A millionaire's son can afford to be a bum.

What must happen instead is the identification and development of those technologies which are in harmony with nature. We need technologies that are environmentally sound, nonpolluting, and nondestructive of the ecology of an overcrowded planet. New technologies must replace older, harmful ones, just as new energy sources must replace oil. New consumer products based on the new technologies will be exciting and will change the consumption and culture patterns of both developed and developing nations.

There are many such technologies now within our grasp, and doubtless many more not yet invented. This book is about one of them—electronic communications.

The uses of telecommunications described in this book will change work patterns, leisure time, education, health care, and industry. The news media, the processes of government, and the workings of democracy could be fundamentally improved. The entire texture of society will be changed by telecommunications and related products. The new technology should give a new hope to today's college generation who must mold its use.

Telecommunications facilities can act as a substitute for much travel, with people able to see each other and operate machines at a great distance. They consume little energy compared with travel. Communications satellites generate their own energy from sunlight in space. New optical-fiber cables, semiconductor lasers that work with them, and microminiature electronic circuitry are manufactured with silicon, one of the earth's most abundant elements. Such satellites and fibers could transmit all the information the human race could possibly use.

Whatever the limits to growth in other fields, there are no limits near in telecommunications and electronic technology. There are no limits near in the consumption of information, the growth of culture, or the development of the human mind.

Civilization is made possible by *communication* in its various forms. The level of communication distinguishes humankind from the animals and distinguishes modern societies from primitive societies. Communications media have done much to alter the nature of society. The impact of television, for example, is far greater than most of us realize. It is now possible to enormously improve society's communications facilities.

The technology of communications is in a period of revolutionary change. Many new inventions and developments are involved. The intent of this book is to describe, in nontechnical terms, the capabilities of the new technology and its potential for both developed and developing nations. Building telecommunications channels is expensive, as was the building of highways. Eventually it will have more impact on society than building highways has had. We can ask, Where should the money come from? Highways are built by governments. In some countries all telecommunications facilities are operated by government; in others by private enterprise. The facilities we describe will become essential to society's infrastructure, essential to the way society is governed, essential to its response to the energy crisis, essential to productivity and hence to the wealth-generating processes. Communications media will be the cornerstone of the culture of our time. Is this important enough to mandate government involvement, for example, in the launching of communications satellites?

There are problems other than costs that will impede the development of telecommunications society. It is important to understand the potentials of such technology and to realize what problems must be solved. Young persons in particular should understand how much of their future can be affected by the government actions described in this book.

Societies growing toward new forms of greatness have often had an image of what might become possible, a vision that inspires the young and draws out the best in people. It is vitally important to understand that the immense riches of today's technology permit us to create visions of a better world. It is possible *now* to build a world without pollution, without massive destruction of nature's beauty, without human drudgery, in which destructive consumption patterns are avoided, and in which the human mind can be nourished as never before in history and can soar to new forms of greatness.

Chapter Two

NEW HIGHWAYS

It is desirable that young people especially, who will inherit the world we are creating, should understand the new opportunities and make their voices heard.

NEW HIGHWAYS

For now sits Expectation in the air.

WILLIAM SHAKESPEARE
King Henry V

Imagine a city ten or twenty years in the future, with parks and flowers and lakes, where the air is crystal clear and most cars are kept in large parking lots on the outskirts. The high-rise buildings are not too close, so they all have good views, and everyone living in the city can walk through the gardens or rain-free pedestrian malls to shops, restaurants, or pubs. The city has cabling under the streets and new forms of radio that provide all manner of communication facilities. The television sets, which can pick up many more channels than today's television, can also be used in conjunction with small keyboards to provide a multitude of communication services. The more affluent citizens have 7-foot television screens, or even larger.

There is less need for physical travel than in an earlier era. Banking can be done from the home, and so can as much shopping as is desired. There is good delivery service. Working at home is encouraged and is made easy for some by the videophones that transmit pictures and documents as well as speech. Meetings and symposia can be held with the participants in distant locations.

Some homes have machines that receive transmitted documents. With these machines one can obtain business paperwork, news items selected to match one's interests, financial or stock market reports, mail, bank statements, airline schedules, and so on. Many of these items, however, are best viewed on the home screens rather than in printed form.

There is almost no street robbery, because most persons carry little cash. Restaurants and stores all accept bank cards, which are read by machines and can be used only by their owners. When these cards are used to make payments, funds may be automatically transferred between the requisite bank accounts by telecommunications. Citizens can wear radio devices for automatically calling police or ambulances if they wish. Homes have burglar and fire alarms connected to the police and fire stations.

Industry is to a major extent run by machines. Automated production lines and industrial robots carry out much of the physical work, and data processing systems carry out much of the administrative work. There is almost no machine tool that does not contain a miniature computer. Paperwork is largely avoided by having computers send orders and invoices directly to other computers and by making most payments, including salary payments, by automatic transmission of funds into the appropriate bank accounts. To avoid unemployment, long weekends have become normal and are demanded by the labor unions.

Inventing and producing ways to fill the increased leisure time is a major growth industry. It cannot be done by driving aimlessly around in an automobile, because petroleum has quadrupled in price since the 1970s. There are sailboats, sailplanes, windjammer cruises, high-fidelity television, and elaborate hobbies. The home screens can be used for playing games with distant opponents, for transmission of sports events, for requesting new movies, for talking face to face, and for remote attendance at conferences.

Above all there is superlative education. History can be learned with programs as gripping and informative as Alistair Cooke's *America.* University courses modeled on England's *Open University* use television and remote computers; degrees can be obtained via television. Computer-assisted instruction, which was usually crude and unappealing in its early days, has now become highly effective. The student watches color film sequences or reads still frames and is asked to respond periodically on a keyboard. His or her response determines what will be shown next. The computer reinforces the material until it is learned. To prepare such programs, there has grown up an industry as large as Hollywood and just as professional. Program production is expensive, but one program is often used by hundreds of thousands of students. You can learn hobbies, languages, mathematics, cooking—all manner of academic and leisure subjects. The world supply of such programs is rapidly growing, and most can be obtained on request on the home communications facilities. The automated education leaves teachers free to concentrate on the more human and creative aspects of teaching.

Information retrieval facilities give access to sports and financial information, weather forecasts, encyclopedias, and vast stores of reports and documents. However, far from acting as a substitute for books, the screens allow citizens to explore the contents of excellent libraries from which books or magazines can be delivered.

The communication channels provide excellent medical facilities, some computerized and some via the videophones and large television screens. Remote diagnostic studios are used, employing powerful television lenses and many medical instruments. With the help of a nurse in the studio, a distant doctor or specialist can examine a patient as though he or she were in the doctor's office. The patient can see and talk to the doctor, and the doctor's large color screen can be filled with the pupil of a patient's eye, or tongue, or skin rash. The doctor can listen to a distant stethoscope and see multiple instrument readings and computer analyses of them. Automatic monitoring of chronically sick patients is done with radio devices, sometimes with automatic administering of drugs. Patients can be monitored during normal daily activities by means of miniature instrumentation (as astronauts were in the 1970s). Patients in remote areas have "telemedicine" access to highly qualified and specialized doctors and facilities when they need them. Most hospitals have telecommunications access to the world's specialists and specialized computers.

Many other aspects of city life are automated. Machines keep track of theater bookings, dentist appointments, health surveillance, and so on. The systems are entirely flexible: When the city dwellers in their wanton way break appointments, fail to turn up, or change their minds at the last minute, the computers reschedule as well as possible. Many persons have developed an arrogant attitude toward this, considering it their prerogative to change their mind whenever they wish and expecting the computers to adjust accordingly. Many of the mechanisms of the city, then, have become part of an enormous programmed complex of machines. Far from being ruled by the computer, as some people feared in earlier decades, the citizen now expects to go his own sweet way and have the computers serve him, slavelike. When they fail and he is kept waiting in a station or cannot obtain the movie he wants on his home screen, he soon protests that the level of programming or facility planning must be improved.

Technical innovation has changed the news media. Citizens can watch their political representatives in action and can register approval or protest. The electorate are both better informed and better able to make their views or protests known.

In the city the cabling that provides these facilities becomes as important as its water pipes and electric supply. On a global scale another telecommunications revolution has occurred. Communications satellites have increased in power so that the images on the home screens, the signals that control industry, the electronic mail, and messages can pass around the world at low cost. The news of Lincoln's assassination took twelve days to reach London. These new signals pass around the world in a fraction of a second.

If new telecommunications has changed the city, it has changed the rural districts even more. For a century there had been movement of people from the country to the city—leaving the farms as farm mechanization spread, looking for work in the cities, seeking their fortune, sending their children away to where there was better education.

Many people would have preferred the countryside if only it had the facilities of the city—good education, excellent medical facilities, superb entertainment, and, most important, highly paid jobs. *Now* telecommunications provides these. Many country villages have a satellite antenna. People can have their own garden or farmstead and can walk in the fields and woods; they eat fresh vegetables and bread from the local bakery; but they are no longer cut off from the world. Offices in the village street are part of big insurance companies or global corporations. They no longer need to be in big cities. The big city offices have become distributed. Their television screens, meeting studios, electronic mail and secretarial services connect them to the world. They can pay the same salaries as in the cities.

The cost of physical transport has risen appallingly with the worsening petroleum shortage. There is an increasing tendency to consume local produce rather than transport it from distant places. Commuting into cities, an unpleasant and time-wasting occupation, has declined. There is a growing trend to small communities which are self-dependent except for their use of the new telecommunications highways.

Country villages have electronic access to the world's medical specialists and computers. Large numbers of television channels are available via the satellite and cable systems, and the excellent education channels reach town and country alike.

Some of the most brilliant individuals are content to live in the remotest villages, where they find peace and beautiful scenery, because they can commune with distant colleagues in many locations at someone else's expense and can use distant machines to conduct research, make video programs, design corporate facilities, or otherwise earn a living. Not all such people want to be remote. Some need the bustling, turbulent, human pressure cookers of the cities and campuses. In a telecommunications society they have a choice. Some spend part of their year in each environment.

The new village culture plays a part in saving energy. The houses use solar energy and wood stoves, backed up by conventional power. The villagers walk to work and often shop by telecommunications. Not much energy is consumed by the microminiature electronics and communications links. The villages are more self-sufficient than in an earlier age, and more ecologically balanced.

Small is beautiful if the pieces are in communication.

The images on the home screens can come from all over the world. Many television programs are dubbed in multiple languages, and the viewer can select the language he wants. In developing nations hundreds of millions of people, unable to read and isolated until recently from most communications, join the club of world television. This immensely powerful medium enlightens, educates, entertains, spreads literacy, and spreads better farming methods, but it can also misinform, spread unfulfillable aspirations, foster greed, provoke antagonisms, discontent, and spread themes of chaos. Only occasionally does it spread the best of human culture.

Medical assistance can be brought by telecommunications to parts of the world where doctors or specialists are in short supply. The satellite channels also make available the services of other experts: crop disease experts, consulting engineers, world authorities. The cost of air travel has increased greatly, but telecommunications make the world a small place.

Multinational corporations are laced together with worldwide networks for telephones, instant mail, and links between computers. Video conference rooms and computerized information systems increase the degree to which head-office executives guide corporate operations in other countries. Computers schedule fleets and optimize the use of resources on a worldwide basis. Money can be moved electronically from one country to another and switched to different currencies. There is worldwide management of capital inventory control, product design, bulk purchasing, computer software, and so on. Local problem situations can trigger the instant attention of head-office staff.

The growth and efficiency of industry has always been highly dependent on communications channels. The new channels interlinking computers and other machines will have a major effect on industry and on the gross national products of nations.

If a better type of cooker or pocket calculator is invented, it is likely to reach the marketplace. With communication media this is not necessarily so. There are complex laws governing communications, and a new technology may not fit into the multifold regulations that exist.

In the past, lawyers and governments have found tortuous reasons to inhibit improvements in communications media. After the invention of power-driven printing presses in England, the British government imposed a tax that required a one-penny stamp on every published page. In 1830 the first penny newspaper was published in the United States, but in England there was a prohibitive four pennies of tax on such a paper. It was called a "tax on knowledge."

When radio was spreading rapidly in the early 1930s,

the American Newspaper Publishers Association attempted to suppress the broadcasting of news. A ruling known as "the Versailles Treaty" because of its harshness permitted radio stations to purchase news from the wire services only on condition that they broadcast no more than ten minutes of news per day in two five-minute segments and that no news item would be more than thirty words long. Sponsorship of news was not permitted. CBS and NBC withdrew completely from the news collection field. The ruling broke down in 1938.

When the first communications satellites were about to be launched and it was clear that they could revolutionize telecommunications, Congress passed the Communications Satellite Act, which prevented satellites being used for transmission within the United States; they were used only for international transmission. The nation that was to put men on the moon could not use its own satellites to benefit its own economy.

The main reason for such harmful laws is that when new technology is introduced, large organizations committed to an older technology can be hurt. The Pony Express went bankrupt two years after Western Union built the first telegraph system across the West. This pattern is normal in competitive industry. The manufacturers of mechanical watches needed to diversify quickly when electronic watches came to the marketplace. In telecommunications, complete freedom of action is prevented by regulations. Corporations that fear a new technology often put massive pressure on the regulations to inhibit its use. Television broadcasters lobbied to stop the spread of cable television, with some measure of success. Communication authorities with huge investments in terrestrial links have tried to slow the development of satellites.

Kinsley, in a study of communications legislation, uses the phrase "corporate luddites." Luddites were members of an English working class movement in the early nineteenth century who smashed the new labor-saving textile machinery powered by steam engines, because they feared it would cause unemployment. Kinsley's corporate luddites attempt to prevent the spread of new technologies which they fear will harm their profits by competing with the status quo. New technology is not smashed *physically* today; it can be destroyed more permanently by lawyers.

Unless the market forces are quick and powerful, old technology always has a momentum that keeps it going long after it is obsolete. It is difficult for the establishment to accept a change in culture or procedure. Nelson's sailing ships successfully blockaded France for two years before Trafalgar. Napoleon did not realize that a steamship was already operating which could cross the English Channel when there was no wind. In the early days of telephony the *Times* of London stated that the telephone was of little value in England because England "had an adequate supply of messenger boys." Today, urban planners ignore telecommunications as being irrelevant even though telecommunications could completely change commuting patterns. Central business districts with sprawling suburbs and increasing numbers of automobiles still represent the standard urban model in spite of rising petroleum costs and growing pollution.

Many of the new uses of telecommunications described in this book are in conflict with the established order. They will encounter fierce opposition from vested interests. However, most of the changes are changes for the better: better education, better news media, improvements in the political process, better forms of human communication, better entertainment, better medical resources, less pollution, less human drudgery, less use of petroleum, more efficient industry, and a better informed society with a rich texture of information sources.

Most persons, and most politicians, are simply not aware of the possibilities, and hence the vested interests may succeed in preserving older methods. It is desirable that young people especially, who will inherit the world we are creating, should understand the new opportunities and make their voices heard.

Chapter Three

MEDICAL FACILITIES

Are the world's telecommunications administrators changing to meet the new social responsibilities?

A man of twenty-five is driving fast along the New York State Thruway shortly after midnight, listening to soporific night-club music. Snow is flurrying in the headlights. Unexpectedly, his car hits a patch of black ice. It spins out of control. He struggles to correct it, but it skids off the road and crashes down an embankment out of sight. Nobody sees the accident.

He is critically injured. He must be taken to a properly equipped hospital quickly. If this accident had happened in the early 1970s, he would certainly have died.

Now, however, a red light flashes in a police patrol car. The policeman checks his terminal, obtains the exact location of the accident, and speeds immediately to it. The man's wallet contains a magnetic-stripe bank card. The policeman finds this and inserts it into the terminal in the police car. He radios an approximate description of the injuries. The victim's medical history record is checked by computer and it is found that he is allergic to certain drugs. His blood group is checked, and the correct blood ordered. His health insurance is also checked, and an appropriate hospital is selected. Instructions to the ambulance staff spatter out on a small printer in the ambulance. The ambulance has automated equipment to keep the victim alive until he reaches the hospital. When he arrives, an operating room is waiting for him. The staff have checked his electrocardiogram report, transmitted from the ambulance and various medical records. A few weeks later he has fully recovered.

Better medical care is one of the many applications of the new telecommunications links that can now be built. Doctors and specialists are in short supply and cannot be everywhere they are needed. Not only doctors, the equipment used in hospitals is becoming more and more expensive. Often it is so costly that only a well equipped hospital can afford it, certainly not a nursing home or a doctor's office. Telecommunications can sometimes make both equipment and doctors usable at many distant locations.

Remote rural communications and small towns are particularly in need of medical access. Sometimes even locations in busy cities are isolated from the medical facilities they need. For example, Logan Airport in Boston is the eighth busiest in the world. Many thousands of passengers pour through the airport all day, and in addition there are more than 5,000 airport employees. Many people are sick at airports, but like most airports Logan does not have a physician in residence. Many such locations are too small to support a full-time physician but too large to be ignored, and a physician generally cannot afford to make journeys back and forth to them. Logan is effectively isolated for many hours of the day by rush-hour traffic congestion.

Logan Airport solved this problem in a pioneering way. It has a medical diagnosis studio connected by a television link to the Massachusetts General Hospital. Many thousands of people per year have had their pains and problems diagnosed via this link. Social scientists' surveys have indicated that most patients are favorably impressed with this way of visiting a doctor,[1] and the nurses present comment that the patients like it. Some patients even say they prefer television visits to being in the doctor's physical presence.[2]

Crichton[3] described a typical patient's use of the system:

> Flight 404 from Los Angeles to Boston was somewhere over eastern Ohio when Mrs. Sylvia Thompson, a fifty-six-year-old mother of three, began to experience chest pain.
>
> The pain was not severe, but it was persistent. After the aircraft landed, she asked an airline official if there was a doctor at the airport. He directed her to the Logan Airport Medical Station, at Gate 23, near the Eastern Airlines terminal.
>
> Entering the waiting area, Mrs. Thompson told the secretary that she would like to see a doctor.
>
> A nurse came over and checked her blood pressure, pulse, and temperature, writing the information down on a slip of paper.
>
> The door to the room nearest Mrs. Thompson was closed. From inside, she heard muffled voices. After several minutes, a stewardess came out and closed the door behind her. The stewardess arranged her next appointment with the secretary, and left.
>
> The secretary turned to Mrs. Thompson. "The doctor will talk with you now," she said, and led Mrs. Thompson into the room that the stewardess had just left.

It was pleasantly furnished with drapes and a carpet. There was an examining table and a chair; both faced a television console. Beneath the TV screen was a remote-control television camera. Over in another corner of the room was a portable camera on a rolling tripod. In still another corner, near the examining couch, was a large instrument console with gauges and dials.

"You'll be speaking with Dr. Murphy," the secretary said.

A nurse then came into the room and motioned Mrs. Thompson to take a seat. Mrs. Thompson looked uncertainly at all the equipment. On the screen, Dr. Raymond Murphy was looking down at some papers on his desk.

The nurse said: "Dr. Murphy."

Dr. Murphy looked up. The television camera beneath the TV screen made a grinding noise, and pivoted around to train on the nurse.

"Yes?"

"This is Mrs. Thompson from Los Angeles. She is a passenger, fifty-six-years old, and she has chest pain. Her blood pressure is 120/80, her pulse is 78, and her temperature is 101.4."

Dr. Murphy nodded. "How do you do, Mrs. Thompson."

Mrs. Thompson was slightly flustered. She turned to the nurse. "What do I do?"

"Just talk to him. He can see you through that camera there, and hear you through that microphone." She pointed to the microphone suspended from the ceiling.

"But where is he?"

"I'm at the Massachusetts General Hospital," Dr. Murphy said. "When did you first get this pain?"

"Today, about two hours ago."

"In flight?"

"Yes."

"What were you doing when it began?"

"Eating lunch. It's continued since then."

"Can you describe it for me?"

"It's not very strong, but it's sharp. In the left side of my chest. Over here," she said pointing. Then she caught herself, and looked questioningly at the nurse.

"I see," Dr. Murphy said. "Does the pain go anywhere? Does it move around?"

"No."

"Do you have pain in your stomach, or in your teeth, or in either of your arms?"

"No."

"Does anything make it worse or better?"

"It hurts when I take a deep breath."

"Have you ever had it before?"

"No. This is the first time."

"Have you ever had any trouble with your heart or lungs before?"

She said she had not. The interview continued for several minutes more, while Dr. Murphy determined that she had no striking symptoms of cardiac disease, that she smoked a pack of cigarettes a day, and that she had a chronic unproductive cough.

He then said, "I'd like you to sit on the couch, please. The nurse will help you disrobe."

Mrs. Thompson moved from the chair to the couch. The remote-control camera whirred mechanically as it followed her. The nurse helped Mrs. Thompson undress. Then Dr. Murphy said: "Would you point to where the pain is, please?"

Mrs. Thompson pointed to the lower-left chest wall, her finger describing an arc along the ribs.

"All right. I'm going to listen to your lungs and heart now."

The nurse stepped to the large instrument console and began flicking switches. She then applied a small, round metal stethoscope to Mrs. Thompson's chest. On the TV screen, Mrs. Thompson saw Dr. Murphy place a stethoscope in his ears.

"Just breathe easily with your mouth open," Dr. Murphy said.

For some minutes he listened to breath sounds, directing the nurse where to move the stethoscope. He then asked Mrs. Thompson to say "ninety-nine" over and over, while the stethoscope was moved. At length he shifted his attention to the heart.

"Now I'd like you to lie down on the couch," Dr. Murphy said, and directed that the stethoscope be removed. To the nurse: "Put the remote camera on Mrs. Thompson's face. Use a close-up lens."

"An eleven hundred?" the nurse asked.

"An eleven hundred will be fine."

The nurse wheeled the remote camera over from the corner of the room and trained it on Mrs. Thompson's face. In the meantime, Dr. Murphy adjusted his own camera so that it was looking at her abdomen.

"Mrs. Thompson," Dr. Murphy said, "I'll be watching both your face and your stomach as the nurse palpates your abdomen. Just relax now."

He then directed the nurse, who felt different areas of the abdomen. None was tender.

"I'd like to look at the feet now," Dr. Murphy said. With the help of the nurse, he checked them for edema. Then he looked at the neck veins.

"Mrs. Thompson, we're going to take a cardiogram now."

The proper leads were attached to the patient. On the TV screen, she watched Dr. Murphy turn to one side and look at a thin strip of paper.

The nurse said: "The cardiogram is transmitted directly to him."

"Oh my," Mrs. Thompson said.

While the examination was proceeding, another nurse was preparing samples of Mrs. Thompson's blood and urine in a laboratory down the hall. She placed the samples under a microscope attached to a TV camera. Watching on a monitor, she could see the image that was being transmitted to Dr. Murphy. She could also talk directly with him, moving the slide about as he instructed.

Mrs. Thompson had a white count of 18,000. Dr. Murphy could clearly see an increase in the different kinds of white cells. He could also see that the urine was clear, with no evidence of infection.

Back in the examining room, Dr. Murphy said: "Mrs. Thompson, it looks like you have a pneumonia. We'd like you to come into the hospital for X rays and further evaluation. I'm going to give you something to make you a little more comfortable."

He directed the nurse to write a prescription. She then carried it over to the telewriter, above the equipment console. Using the telewriter unit, Dr. Murphy signed the prescription.

Afterward, Mrs. Thompson said: "My goodness. It was just like the real thing."

The doctor in the studio at Massachusetts General Hospital can see and adjust his own image on a monitor screen and can check his facial expressions. He operates the camera in the distant examining room by means of a joystick. When he pushes this up, down, left or right, the camera moves the same way. He has a focusing control and can zoom from a view of the whole room to a close-up of the patient. There are two other cameras, one which the nurse in the examining room sets up and the other in the laboratory. Microscope examinations can be transmitted from the laboratory to the doctor. X-ray inspection has also been done via telecommunications.

Nobody would claim that *all* diagnoses can be handled over the television link. Some attempted diagnoses result in the need for the doctor and patient to be physically present. However, this was considered exceptional in the Logan Airport experience.

SPECIAL OPINIONS

In addition to doctor–patient telecommunications, video links can permit physicians to confer. A doctor examining a patient and wanting a specialized opinion could contact a colleague via television and switch any of the cameras to his screen. The doctors could discuss the case, either within the patient's hearing or not.

Hospitals of the future may have switched television links interconnecting doctors' offices, examining rooms, patient bedsides, laboratories, and operating rooms. In Vermont, New Hampshire, and Maine a private two-way television network interconnects many hospitals. A doctor in a small hospital may confer with experts or use machines that are available only in the large or specialized hospitals. Authorities giving lectures can be seen anywhere on the network.

Many lectures on specialized aspects of medicine are video taped. There are many thousands of such tapes, and eventually there will be hundreds of thousands. It would be helpful for doctors or medical staff to be able to see such tapes, either when a patient is involved or for general education. It has been suggested that a library of tapes on all manner of medical conditions should be produced *for patient viewing*. This could explain to a patient the nature of his condition in a more thorough manner than his doctor's explanation, which is often minimal. It could mold a patient's actions and attitudes so that he would help himself recover.

COMPUTER DIAGNOSIS

A particularly intriguing experiment in medicine has been the use of computers to interview patients. The computer can collect the type of information that is needed in medical diagnosis. It can tell a patient or nurse whether the patient should see a physician, and what type of physician. Although the computer is not likely in the foreseeable future to perform complete medical diagnoses, working systems have shown that it can save much of a physician's valuable time.

The following is a dialogue between a patient and a computer terminal, which has been used for some years at a clinic in Edinburgh concerned with stomach ailments. All new patients irrespective of age or sex are asked to use the terminal. They could reply to the computer's questions with one of three keys, labeled "YES," "NO," and "?"[4]

COMPUTER: THIS IS A COMPUTER TALKING TO YOU. I CAN ASK YOU QUESTIONS BUT I CAN ONLY UNDERSTAND SIMPLE ANSWERS, SO YOU CAN ONLY TALK TO ME BY PUSHING EITHER THE **YES**, **NO** or **?** BUTTON. YOU SHOULD ONLY PRESS THE '?' BUTTON IF YOU DON'T UNDERSTAND THE QUESTION I AM ASKING YOU. NOW, DO YOU UNDERSTAND EVERYTHING I'VE SAID UP TO NOW? PUSH ONE OF THE BUTTONS.

PATIENT: YES

COMPUTER: THANKS. NOW I WANT TO ASK YOU ONE OR TWO QUESTIONS ABOUT WHY YOU ARE HERE. PLEASE REMEMBER THAT THIS IS ALL IN COMPLETE CONFIDENCE JUST AS WITH DOCTORS. ONLY DOCTORS AND NURSES WILL SEE YOUR ANSWERS. ARE YOU QUITE HAPPY ABOUT THIS? PUSH ONE OF THE BUTTONS.

PATIENT: YES

COMPUTER: GOOD. BY THE WAY, DON'T FEEL THAT YOU HAVE TO PRESS THE BUTTON THE VERY INSTANT THAT I HAVE FINISHED TYPING. IF YOU FEEL YOU NEED TO THINK A BIT BEFORE YOU ANSWER, OR NEED TO READ THE QUESTION OVER AGAIN THEN GO AHEAD AND DO SO. TAKE YOUR TIME IF YOU WANT TO. REMEMBER, HOWEVER, THAT I CAN'T TALK TO YOU AGAIN UNTIL YOU HAVE PUSHED ONE OF THE BUTTONS.
WELL, FIRST OF ALL THEN, IS ONE OF THE REASONS YOU VISITED YOUR DOCTOR BECAUSE YOU WERE SUFFERING FROM DISCOMFORT OR PAINS IN THE STOMACH?

PATIENT: YES

COMPUTER: COULD YOU TELL ME HAS THE PAIN OR DISCOMFORT BEEN CONTINUOUS SINCE IT STARTED? PRESS ONE OF THE BUTTONS.

PATIENT: NO

COMPUTER: I'D JUST LIKE TO BE QUITE CERTAIN OF THIS POINT. HAVE YOU HAD TIMES, DAYS OR WEEKS, WHEN YOU'VE BEEN COMPLETELY FREE OF THE PAIN?

PATIENT: YES

COMPUTER: I SEE. TELL ME IS THE DISCOMFORT RELATED TO TAKING FOOD—DOES EATING HAVE SOMETHING TO DO WITH IT? PUSH "YES" OR "NO".

PATIENT: ?

COMPUTER: I AM TRYING TO FIND OUT WHETHER THIS PAIN OR DISCOMFORT OF YOURS IS IN SOME WAY AFFECTED BY THE THINGS THAT YOU EAT OR WHEN YOU EAT. IF IT DOES SEEM TO HAVE SOMETHING TO DO WITH EATING PRESS "YES," IF NOT PRESS "NO."

PATIENT: YES

COMPUTER: WHEN YOU HAVE THE PAIN OR DISCOMFORT DO YOU GET RELIEF FROM INDIGESTION TABLETS OR LIQUID MEDICINES? PRESS "YES" OR "NO."

PATIENT: YES

COMPUTER: TELL ME DOES THE PAIN EVER WAKE YOU IN THE EARLY HOURS OF THE MORNING? "YES" OR "NO"?

PATIENT: NO

COMPUTER: NOW I'D LIKE YOU TO THINK RATHER CAREFULLY BEFORE YOU ANSWER THE NEXT QUESTION. JUST IMAGINE THAT I ASK YOU TO POINT OUT THE AREA WHERE YOU FEEL THE PAIN IN YOUR STOMACH—NOW CAN THE AREA BEST BE INDICATED WITH ONE OR TWO FINGERS, OR WITH THE FLAT OF THE HAND.
IF WITH ONE OR TWO FINGERS, PRESS "YES"
IF WITH THE FLAT OF THE HAND, PRESS "NO"
IF YOU DON'T UNDERSTAND, PRESS "?"

PATIENT: NO

COMPUTER: I EXPECT YOU KNOW WHERE THE NAVEL IS, BUT I WANT TO BE QUITE SURE ABOUT THIS. PUSH "YES" IF YOU KNOW WHERE THE NAVEL IS AND "NO" IF YOU DON'T KNOW.

PATIENT: YES

COMPUTER: OK. NOW IS THE PAIN OR DISCOMFORT, WHEN YOU GET IT, MAINLY ABOVE THE NAVEL? THINK CAREFULLY AND PUSH "YES" OR "NO."

PATIENT: YES

COMPUTER: THANK YOU. I NOW NEED TO FIND OUT WHETHER THE PAIN OR DISCOMFORT IS TO THE LEFT OR RIGHT OF YOUR STOMACH, OR WHETHER IT LIES GENERALLY IN THE CENTRE (WHICH WE CALL THE MIDLINE), AND I WILL ASK YOU ONE OR TWO QUESTIONS ABOUT THIS. FIRST OF ALL DOES IT LIE ON THE MIDLINE? PUSH "YES" IF IT DOES, AND "NO" IF IT DOESN'T.

PATIENT: YES

COMPUTER: WOULD YOU SAY THAT YOU WERE EATING NORMALLY THESE DAYS? BY NORMALLY, I MEAN DO YOU HAVE A REASONABLY GOOD APPETITE? PRESS ONE BUTTON NOW.

PATIENT: YES

COMPUTER: WELL YOU ARE BEING VERY HELPFUL SO FAR AND I WOULD NOW LIKE TO ASK SOME MORE QUESTIONS. I HOPE YOU STILL FIND THIS INTERESTING. IF YOU DO PUSH THE "YES" BUTTON, BUT IF YOU ARE GETTING BORED, PUSH THE "NO" BUTTON.

PATIENT: YES

COMPUTER: IN THIS PRESENT ILLNESS HAVE YOU HAD ANY VOMITING? PUSH "YES" IF YOU HAVE AND "NO" IF NOT.

PATIENT: YES

COMPUTER: COULD YOU TELL ME, WHEN YOU VOMIT DO YOU BRING UP FOOD?

PATIENT: NO

COMPUTER: DO YOU SOMETIMES FIND THAT YOUR MOUTH UNEXPECTEDLY FILLS UP WITH A FLUID? PUSH "YES" OR "NO" PLEASE.

PATIENT: YES

COMPUTER: IF YOU HAD TO DESCRIBE THIS LIQUID OR FLUID WOULD YOU SAY IT WAS CLEAR AND TASTED SALTY?

PATIENT: YES

COMPUTER: DOES THE FLUID EVER HAVE A BITTER OR ACID TASTE? PRESS "YES" IF IT DOES, AND "NO" IF NOT.

PATIENT: YES

COMPUTER: THANK YOU, WE ARE GETTING ON VERY WELL AND I HAVE ONLY ONE OR TWO MORE QUESTIONS TO ASK YOU. NOW I EXPECT YOU KNOW WHAT HEARTBURN IS. IT IS PERHAPS BEST DESCRIBED AS A BURNING FEELING, BEHIND THE LOWER END OF YOUR BREASTBONE. TELL ME, DO YOU SUFFER FROM HEARTBURN A LOT?

PATIENT: NO

COMPUTER: DO YOU TEND TO GET AN ACID FLUID COMING INTO YOUR MOUTH AFTER EATING FOOD? "YES" OR "NO"?

PATIENT: NO

COMPUTER: VERY WELL, DO YOU TEND TO GET AN ACID FLUID IN YOUR THROAT, WHEN YOU LIE DOWN AT NIGHT OR WHEN YOU CHANGE YOUR BODY POSITION, SUCH AS STOOPING, GARDENING, LIFTING SOMETHING HEAVY? IF YOU DO PUSH THE "YES" BUTTON, IF NOT PUSH THE "NO" BUTTON.

PATIENT: NO

COMPUTER: NOW A QUESTION ABOUT YOUR WEIGHT. IN THE LAST SIX MONTHS HAVE YOU LOST RATHER A LOT OF WEIGHT, SAY OVER 10 POUNDS, WITHOUT DELIBERATELY SLIMMING? PRESS "YES" IF YOU HAVE, AND "NO" IF NOT.

PATIENT: NO

COMPUTER: ONE FINAL QUESTION, WHICH I WOULD LIKE YOU TO THINK CAREFULLY ABOUT. HAVE ANY MEMBERS OF YOUR FAMILY—YOUR FATHER, MOTHER, UNCLES, AUNTS, BROTHERS OR SISTERS—EVER SUFFERED FROM STOMACH OR DUODENAL ULCERS AS FAR AS YOU KNOW? IF ANY OF THEM HAVE HAD AN ULCER, PRESS "YES," IF THEY HAVEN'T PRESS "NO."

PATIENT: NO

COMPUTER: WELL, THANK YOU VERY MUCH. THAT'S ALL I WANT TO ASK YOU. THANKS FOR BEING SO HELPFUL. IN A SECOND THE PAPER WILL START TO UNROLL AND WHEN IT STOPS PLEASE TEAR IT OFF AND HAND IT TO THE NURSE OUTSIDE. GOODBYE.

The ease with which the patients accepted the computer's questioning is surprising. There were few, if any, signs of anxiety or uncertainty. Questioned about the experience afterward, none of the patients said that they found it unpleasant or annoying. Many compared the computer favorably to the live consultant, describing it as "polite," "friendly," and "understandable." In one similar application, designed to establish a personality profile of the patient, it was even difficult to persuade old ladies to leave the terminal; nobody had taken that much interest in them in years. In some cases patients have expressed a strong preference for communicating with a terminal to communicating with a live person. This is especially true with patients who are embarrassed about their ailments or have difficulty in human communication. With some psychiatric dialogues, especially about sexual problems, computer interviews have succeeded in collecting information which a doctor could not.

The above dialogue was being used to collect *indicants*—a medical term for pieces of evidence about whether a disease is present. The performance of the system was investigated by having physicians perform a diagnosis on the patients who used the terminal. The investigating team concluded that the group of indicants in question were "elicited by the computer with approximately the same precision as that from an expert consultant.[5] Some sets of indicants can reveal that the person does not appear to need a doctor; an Alka Seltzer would suffice. Others indicate that medical attention is definitely needed, and they might direct the patient to a particular specialist. The more complex the indicants used or the greater the range of indicants searched for, the more thorough will be this prediagnostic information. As medicine becomes more formalized, computers will be used to a greater extent to assist in the diagnostic process.

With telecommunications a dialogue like the above could easily be carried out *from the home.* The computer could display questions and possibly diagrams on the home television screen, and the user would respond. This could become a valuable service. A person feeling ill might dial a medical computer. The computer would interview the person and might decide whether to switch the person to communication with a physician. Some persons might insist on talking to the physician regardless of the computer—but that would be more expensive.

MEDICAL RECORDS

The above process, and indeed any medical diagnosis, is likely to be more thorough if the patient's medical history is taken into consideration. Currently the medical records of most people, if they have been kept at all, reside in a cardboard folder in their doctor's office. Some doctors and some clinics have many shelves full of such folders recording patients' past ailments. When people move, their records more often than not do not move with them. They may have many sets of records, none giving a complete medical history, scattered among different doctors.

Increasingly, medical records are being kept by computers, which can lessen the total cost. This has several advantages. The records are likely to be more thorough, precise, and methodical. They are unlikely to be lost. Particularly important, because the computers are connected to telecommunication links, the records can be made accessible anywhere they are needed. A doctor with a patient in his office should be able to call up the patient's history on his screen and add items to it if necessary.

A SCREEN WITH MANY FUNCTIONS

A physician with good telecommunications would use his television-like screen for multiple purposes—interviewing patients, inspecting laboratory results, obtaining a computer's prediagnosis results, obtaining patient records, communicating with specialists or obtaining second opinions, referring to encyclopedic data banks of medical information, and on-going personal education.

The role of nurses and "paramedical" staff has been steadily increasing. With better telecommunications facilities, these staff may supervise the use of computer terminals, television clinics, electrocardiogram machines, and so on. They may recommend whether a patient should see a doctor, and possibly which doctor. Paramedics may become the major agents dispensing routine care and could do this much better if they have excellent telecommunications links to physicians. The paramedics may take over more of the "samaritan" role while physicians increase their involvement in the scientific role.

The role of electronics in medicine is rapidly increasing, and as it increases, better telecommunications facilities are needed.

VILLAGES WITH SATELLITE ANTENNAS

Amazingly, 134 counties in the United States are without a single practicing non-Federal physician.[6] Doctors often have urban tastes or like to be near large hospitals, and they tend to cluster in urban areas. Furthermore, the number of doctors in any area in the United States varies greatly with the average income of the area.[7] In areas with income equal to the national average there are 200 physicians for every 100,000 people; in areas half that wealthy there are only 40 physicians per 100,000 people. Inhabitants of a rural area with low average income can thus be far from medical help, even though America spends vastly more money per capita on medicine than most nations. In developing nations the situation is usually much worse.

Alaska is six times the size of Britain but does not have a medical school. NASA chose Alaska to experiment with health care services via its ATS-6 communications satellite in 1974. The experiment interconnected five sites with links which transmit and receive television, voice, and signals such as electrocardiogram and electroencephalogram output. The earth station equipment which the sites needed was designed to be low in cost and easy to maintain by nonexperts. It used a ten-foot dish-shaped antenna which could be set up and adjusted almost as easily as a television antenna.[8]

Two examining rooms were used in small clinics at Fort Yukon (population 630) and Galena (population 425). Physicians were employed at a Public Health Service Hospital hundreds of miles away in Tanana. These doctors would diagnose patients' problems and often prescribe drugs or therapy which were administered by local medical staff. The doctors would sometimes consult specialists who were at Fairbanks or Anchorage. The specialists could see the patients, laboratory tests, or the consulting doctors. Patient records were obtained, also via satellite, from a computer programmed to handle patient records several thousand miles away in Tucson, Arizona. After the examination the doctors would update these records.

For the sake of privacy all transmissions associated with the consultations were scrambled and could be unscrambled only at the appropriate clinics or physicians' offices.

In general the experiment was highly successful. Several lives were saved in emergency situations. The quality of available medical care was greatly improved. The same satellite was also used for education, television broadcasting, data transmission, and navigational experiments. Large commercial (as opposed to experimental) satellites could give nationwide coverage and could be connected to local television channels such as cable television.

Telecommunications applied to fields other than medicine can also have a dramatic effect. In some fields it will bring changes more pervasive than those in medicine. However, we have to ask questions like: Are the right satellites going to be launched? Who will launch them? Are the world's telecommunications administrations changing to meet the new social responsibilities?

Chapter Four

WIDELY DIFFERING REQUIREMENTS

Humankind is poised midway between the beasts and godlike powers of communication.

A science fiction writer's vision of future telecommunications might assume complete interlinking between rooms where people live and work. The rooms have wall screens that pick up television but are also interconnectable, like today's telephones. When you dial a call, instead of merely hearing a telephone voice at the other end, the wall "dissolves" and you are connected to another apartment or office, anywhere in the world.

Technology has evolved to the stage where this is *possible.* However, the reality of telecommunications is different and more complex, and its potentials are in many ways more interesting. We need to discuss how society can evolve from today's telephone and broadcasting facilities to forms of telecommunications of more value.

LARGE AND SMALL PIPES

Just as plumbing uses pipes of different capacities, so also we have telecommunications channels of different capacities. In fact, there is a much wider range of channel capacities than of pipe sizes. A color television set, for example, needs a transmission channel with almost one million times the capacity needed for a telex machine.

It is the author's intention to avoid technical terms and technical discussions whenever possible. However, the economics of different channel capacities are critical to the arguments about how telecommunications should be used. This chapter therefore gives a brief summary of the different channel types and capacities.

BITS

Information of any type can be transmitted over telecommunications media in two basically different ways: It can be analog or digital.

Analog transmission means that a continuous jumble of frequencies is transmitted. Sound and light both consist of a collection of frequencies. A high-fidelity enthusiast strives to reproduce accurately a continuous range of sound frequencies from very low notes to very high notes. He wants sounds from 30 to 15,000 or (for people with very good ears) 20,000 cycles per second. The frequencies near 20,000 will probably be heard only by the passing bats. If you want to listen to high-fidelity music via the telephone wires into your home (which is possible with certain modifications), a continuous range of frequencies from 30 to 20,000 cycles would be transmitted. The current on the wire would vary continuously in the same way as the sound you hear.

In *digital transmission,* on the other hand, a stream of pulses is sent. Each pulse is in one of two conditions: *On* or *Off,* or perhaps *Positive* or *Negative.* The reader might think of a pipe with tiny balls traveling down it; every ball is identical except that some are *Black* and some are *White.* The pulses, or balls in this analogy, are referred to as *bits.* Any information can be represented as a stream of bits, and this is the way computers represent information. The pipe may have only fifty balls passing any one point in a second, or, on the other hand, it may be a high-capacity pipe with 50 million balls passing one point in a second.

Any information, such as text, telephone speech, music, or pictures, can be transmitted in either an analog form or a digital form. The "pipes" themselves can be designed as either analog channels or digital channels. This applies to all types of transmission paths: wire pairs, high-capacity cables, radio links, satellites, and entirely new transmission media. If the path is designed to be analog, it will use amplifiers somewhat similar to that in a high-fidelity unit for increasing the signal strength. If it is digital, it will use repeaters to reconstruct the bit stream and pass it on.

Channels that transmit streams of bits have come into use for carrying some telephone signals, even though telephone speech is basically analog in nature. Computers and microminiature computer-like circuitry manipulate streams of bits. For technical reasons many signals will be carried as streams of bits in the future, and telephone networks will evolve to being largely digital in form. For an analog signal to be handled in digital form, it must be converted to a stream of bits by a small-mass-produced circuit.

WIDELY DIFFERING REQUIREMENTS

To compare the channel capacities needed for various signals, we list how many bits per second are used today to carry them.

	Bits per Second
Telex transmission (dial-up telegraph machines)	55
Transmission to a medium-speed computer terminal	4,800
Telephone speech	64,000
High-fidelity music	400,000
Picturephone	6.3 million
Color television	92 million

Clearly there is a wide range of channel capacity requirements. Telex transmission requires only a small fraction of a digital speech channel. Television requires one thousand times the capacity of speech. In general the human eye absorbs about one thousand times as much as the human ear, and a large proportion of the brain cells of humans and animals are employed in processing the visual images received. High-fidelity sound needs many more bits per second than telephone-quality sound. Similarly, visual images of widely varying quality can be used. To transmit a moving image of the quality and size of a cinema screen would require about ten times as many bits per second as color television. On the other hand, a small black and white image like AT&T's Picturephone requires roughly one tenth the capacity of television.

Any physical transmission channel can be made to carry many separate signals if it has enough capacity. Many of the copper wire telephone trunks in North America are designed to carry 1.544 million bits per second. They carry 24 telephone signals, each occupying a bit stream of 64,000 bits per second. The same transmission link could be made to carry a small number of hi-fidelity music signals, or a much larger number of telex signals.

With more complicated encoding speech, television, or Picturephone signals could be carried with a smaller number of bits than today's numbers listed above. Today's numbers serve to illustrate the differences in capacity requirements. To wire a society so that television images could be transmitted between homes as today's telephone signals are would require about a thousand times the channel capacity. It would not be a thousand times as expensive, though. In fact, new types of channels now coming into existence could make the future transmission cost of television similar to today's cost of transmitting speech.[1] However, today's society is wired for speech, and it will cost billions of dollars to rewire it for images.

We refer to channels that transmit visual images, such as television or Picturephone, as *video* channels. Broadcast television and cable television provide video channels. Private video channels are used for security purposes within buildings, and there are a few long-distance video links in government and industry. Satellites will make it possible to have video links between office buildings, schools, or factories at reasonable cost.

Aldous Huxley's *Brave New World*[2] described "feelies," in which sensations of touch were transmitted as well as sight and sound: "There's a love scene on a bearskin rug; they say it's marvellous. Every hair of the bear reproduced. The most amazing tactual effects." Unfortunately, technology is *not* likely to be able to produce "feelies" in the near future.

TYPES OF CHANNELS

Telephone channels transmit continuously in both directions so that a conversation can take place. For some telecommunications, *two-way* transmission is not needed. For others, *continuous* transmission is not needed. Television and radio broadcasting employ one-way transmission. The audience cannot respond (unless there are new facilities). The term "broadcasting" usually implies that there are a small number of transmitters and many receivers.

Sending messages does not require continuous transmission like that needed to transmit speech or television. A brief burst of transmission is adequate. For example, a teletype message typically contains about 3000 bits of data. If this is sent over a digital telephone channel operating at 64,000 bits per second, it requires a burst lasting about one twentieth of a second. Thus, thousands of teletype messages could be sent in the time required for one telephone conversation.

Some of the transmission links in North America transmit 274 million bits per second.[3] Such speeds will become common as advanced technology spreads. Over such a link a teletype message takes little more than one hundred-thousandth of a second. Vast quantities of messages could flash across continents if appropriate switching and control mechanisms existed.

A particularly important form of communication is the use of distant computers. A computer terminal, the means of gaining access to a computer, is now a familiar sight on films and television, in offices, and at airports. It may look like a typewriter or telegraph machine, or it may be a device with a keyboard and television-like screen. Millions of terminals are now in use, involved in applications ranging from police work to hotel booking or stock market information, to engineering design.

A person using a terminal often carries out a "dialogue" with a distant computer. The user sends a message to the computer; the computer responds; the user reacts to the response, and so on. A conversation is taking place. However, it is unlike a telephone conversation in that it consists of a set of brief messages with silence between them. Sometimes the messages are very short; the *average* number of bits per second transmitted for the duration of the dialogue is typically less than ten.[4] However, each message must reach its destination quickly, so a line handling thousands of bits per second is needed.

Ten bits per second is a small fraction of the bit rate of a telephone circuit, especially a digital one operating at 64,000 bits per second. Therefore many separate computer dialogues could share such connections if appropriate control mechanisms existed. Many millions of computer dialogues could share telecommunications highways operating at 274 million bits per second.

A vast number of signals are sent during the transmission of one hour of television. If it were transmitted digitally, more than one hundred thousand million bits would be needed. That is equivalent to all the telegrams Western Union handles throughout the world in a year.

If print or computer data is transmitted, much more information can be transferred than during an equivalent transmission of a telephone conversation. The entire Bible could be transmitted at 64,000 bits per second in the duration of one of my wife's telephone calls. It could be transmitted *one thousand times* over a digital television link in the duration of one television talk show.

In summary, different types of messages vary enormously in their consumption of transmission capacity. The cost of sending text or computer messages could be very low if appropriate mechanisms existed. So far, however, society has spent most of its telecommunications money on telephone and broadcasting networks.

Chapter Five

NEW USES OF TELEVISION

The problem is access.

Television is the most influential communications medium in the history of man.

ROBERT SARNOFF,
then Chairman of R.C.A.

Television is a low-brow medium. This is its social role, and there is no sense in attempting to improve it.

FEDERAL COMMUNICATIONS COMMISSIONER
LOWINGER 1967

There are many new ways in which television sets will be used.

VIEWDATA

Both television sets and telephones exist in vast numbers of homes. When two such ubiquitous technologies are wed, it is likely to be the marriage of the decade. The British Post Office now has plans for this marriage. Electronic circuits have been added to the television set which connect it to the telephone network. The user will have a small keyboard about the size of a pocket calculator. The new service that results from the marriage is called "viewdata."

With a *local* telephone call the owner of a television set with the Viewdata addition can gain access to Post Office computer systems which store information and programs. Millions of pages of data will become available, each designed to be displayed in color on the screen of a television set. The home user can look at news reports, movie listings, weather forecasts, stock market figures, racing results, and so on. Business people can look at business information or data banks of their own organizations. Both will carry out *dialogues* with their television sets which enable them to browse through the vast data files.

Computer programs as well as data will be stored on the Viewdata files. These could be used for doing tax returns, selecting or analyzing stocks, carrying out a medical assistance dialogue like that in Chapter 3, learning languages or mathematics with computer-assisted instructions, playing games, and an infinite number of other purposes some of which are mentioned later.

Viewdata customers will be able to keep their own files, and to leave messages for other Viewdata users. A light will glow on the television set when a message awaits its receiver.

Where will all the data and programs come from? The answer to this is perhaps the most brilliant aspect of the Viewdata scheme. It can come from anywhere. Computer hobbyists or individual entrepreneurs can put data or programs into the Viewdata files and when any of the viewers use it the originators will be paid a small royalty. Similarly, advertising organizations, department stores, government agencies, or big corporations can make their data or programs available. It is, in effect, a new form of publishing. Publishing organizations, large and small, will distribute information via Viewdata. But unlike most forms of publishing the private individual can participate.

Some of the data will be constantly updated, such as news, weather, and stock market data. Some, like encyclopedias, will remain unchanged for years. Humankind has vast quantities of data and computer programs which would be valuable in homes and offices. The problem is gaining easy access to it when it is needed. The Viewdata scheme can provide this at the right price. However, there are many obstacles which lawyers will raise to prevent its adoption in some countries.

TWO CABLES

There are two telecommunications cables going into many homes today: the telephone cable and the television cable. The information-carrying capacity of the television cable is thousands of times greater than that of the telephone cable. Comparing them is like comparing a wide flowing river to a kitchen tap.

In spite of the enormous difference in capacity, the television cable costs much less per home than the telephone wiring. Why? First, it uses newer technology: It uses a coaxial cable, whereas the telephone wiring uses old-fashioned wire pairs. Second, one coaxial cable has branches going into hundreds of homes, whereas the telephone system has *one wire pair for each* home. The telephone cable, however, is connected to switching facilities so that any two subscribers may be interconnected.

Cable television has an even greater potential than Viewdata. Many possible uses of the cable have not yet been exploited. With modern electronics, cable television can be used in ways quite different from those originally intended, and the potentialities are exciting.

THE SPREAD OF CABLE TELEVISION

Cable television was started in 1949 by a local radio dealer in an Appalachian village, who little dreamed what he was beginning. Realizing that the surrounding mountains were spoiling the reception of television signals from Philadelphia, so that he would be unable to sell many sets, he conceived an enterprising scheme: He erected a tall antenna on a suitable mountain where the reception was good and ran coaxial cable from the antenna into the homes of people willing to pay a small fee. In this way television came to an Appalachian village, and a technique came into use that should, if allowed, revolutionize home electronics.

Cable television spread steadily in the 1950s and 1960s, although hampered by government regulation and by lobbying from the big three networks, who feared its competition. Its initial spread in the United States was in rural areas where reception was poor. In Canada it spread in the cities, largely because cable television operators were allowed to import distant signals and hence provide more programs. In 1968 a U.S. Justice Department antitrust decision urged the Federal Communications Commission to allow cable television to develop as a competitive medium with its own program origination and advertising. With much catching up to do, cable television systems were then built in U.S. cities.

The main sales arguments today are the provision of advertisement-free programs (such as sports broadcasts, first-run movies, news, and stock market reports) and improved reception. As the pamphlets mailed to New York apartments said,

> Ghosts, unsteady pictures, weird psychedelic colors and assorted eye-jarring interference have plagued TV reception in this area since commercial television first became a reality . . .
>
> The tall buildings between your set and the broadcasting tower deflect the local TV signals. The signals bounce off these buildings before reaching your antenna. The result: ghosts. And that's just the beginning.
>
> TV signals ricocheting off planes flying overhead are subjected to electrical interference ranging from automobile ignitions to your neighbor's vacuum cleaner. They finally reach your set much the worse for these interferences. You get way-out colors, flip-flopping pictures, snow, herringbone patterns, and assorted other "stray" signals in addition to the signal sent out by the station.
>
> . . . As a cable TV subscriber, you, too, can discover the pleasure of clear, sharp, ghost-free, snowless, jitterless television in brilliant color or crisp black and white.[1]

Cable television serves more than 10 million homes in the United States at the time of writing, and the number is increasing rapidly. It has been estimated that by the early 1980s half the homes in North America will be wired for it. There is a wide divergence of opinion about the rate at which it will spread as well as its eventual penetration.

MANY CHANNELS

One cable of reasonable cost can carry forty television channels. More expensive facilities could carry hundreds. Many channels are being used by the Warner Brothers' Qube cable system in a test program in Columbus, Ohio.

The public's appetite for television is enormous. An average American home has the television switched on for 6½ hours per day. Given the vast audience, a question of major social importance is, How should the numerous channels be used? Not all of them need be used for conventional television. It is in alternative uses of the cable that the most interesting potentialities lie.

HOW SHOULD THE CHANNELS BE USED?

Ours is an age dominated not by culture but by technology and lawyers, and the burgeoning cable television industry has difficulty finding ways to fill its many channels.

The most profitable use of the new channel has been for broadcasting popular movies to subscribers for an extra charge.

Some cable systems are using some of their channels for services such as weather reports, news headlines, time signals, and stock market reports. Most of these services are provided very inexpensively with a single fixed camera. Some cable television companies run their own films and produce their own programs, in some cases serving a local rather than a statewide need. Local sports events, local shopping information, politicians whose constituencies number only a few thousand, and so on—these seem natural for cable television.

A television program need not be expensive. Some cable television companies have a very simple studio with little equipment other than lights. The studio can be used at little cost by any group who wants it. A local school board has used this facility to discuss matters such as upcoming bond issue referendums. Only a simple camera is needed for many local events, and a person can learn to operate such a camera very quickly. City council and board meetings, school events, local sports, and amateur entertainment productions can all be shown on the local cable. One system in Manhattan transmits Columbia University's basketball games. Thus, cable television provides a local medium as well as McLuhan's "global" one.

Locally originated television from bargain-basement studios was heralded by some groups with immense enthusiasm. They talked about "minority group programming," "releasing the creativity of the common man," and "escaping the tyranny of the big networks." After a few years of experimentation locally-originated television has varied enormously from one location to another. An amateur with a camera does not necessarily create good television. Regulations mandate that most American cable systems have a *public access channel,* on which any citizen may telecast free of charge, and uncensored. In a few cities this channel vibrates with a cultural diversity foreign to conventional television. In Manhattan it slowly became a medium for the visual arts community carrying anything from night-club shows to poetry, and being used by ministers, masseurs, animal lovers, politicians, and would-be social commentators. In some American cities the uncensored public access channel was used mainly for pornography (greatly increasing the numbers of subscribers).

Future historians may find it surprising that for its first three decades the most powerful communications medium in history was so little used for the improvement of society. There is remarkably little education on television. Pioneering, and otherwise successful, attempts at education such as Britain's Open University have not been given their own channel. Government and community debates are not normally broadcast. We tolerate a road accident rate worse than the casualty rate in most wars; television could be a useful means of showing people the causes of accidents and inculcating safer driving habits. It could teach better nourishment and health habits. The religious leaders of most countries make almost no use of television, and the social attitudes which television repetitively illustrates and glamorizes are mostly harmful ones.

With more channels there would be more scope for socially valuable uses of television.

On a worldwide basis, there are enormous quantities of excellent program material that have not found their way onto the cables. The BBC archives, for example, are gigantic and filled with past television programs on all manner of subjects, too good to deserve a fate of permanently gathering dust. Many countries make good children's programs to which soundtracks could be added in different languages. Japan spends much money on television production. The world would surely be a richer place if the spare television channels could be filled with quality programming from different nations.

Our capability to make films and television gives us a cultural storehouse that is unique in history. The total supply of good films will steadily increase. As with books, the bad and mediocre films will be forgotten, while films of lasting value are steadily added to the store. The film libraries of the world will continually become richer in content. The BBC spends a small portion of its budget attempting to produce high quality dramatizations of the classics. Unlike much of the pulp shown on daily television, these will be of value generations hence. We can capture the great stars, the great performances, the work of great directors, for posterity.

The problem again is *access*. How do we make the gems hidden in the world's film libraries available? With music, the public buys records. It might be worthwhile for the public to buy video records as well, but this does not solve the problem. With few exceptions people watch a film or play only once, whereas they listen to a favorite piece of music many times and like to have it available.

Cable television can help in two ways. First, with many channels programs of high quality can be rebroadcast, and viewers can be given a much wider range of programs to select from. It is much cheaper to broadcast old programs of high quality than to make new ones of low quality. Second, mechanisms can be created for enabling viewers to *select* any film they want to see from a film library. To give a home *complete* freedom in selection, that home would need its own video channel, unshared with other homes (just as it has its own telephone channel today). This is done with some schools, giving the school freedom to request programs from a manually operated library. Although ordinary homes will not have an unshared video channel in the near future, a limited choice of programs, possibly requested by the viewers, could be made available on a group of channels for which the viewers pay a periodic fee—an extension of today's Home Box Office service in the United States.

DIVIDING UP THE CHANNELS

Because a television channel has a high information-carrying capacity—one thousand times that of a telephone channel—any of the channels carried by the television cable could be subdivided into smaller channels. These could be used for speech, high-fidelity music, still pictures on the screen, data transmission, home newspaper delivery, electronic mail, and so on. Many proposals for future domestic products have arisen from these possibilities.

One television channel could carry 100 high-fidelity stereophonic music channels. Out of the many television channels, perhaps one could be divided up for classical and one for pop music. The classical music channel could broadcast 100 pieces (symphonies, operas, and concertos) simultaneously, each occupying a different frequency slot within one television channel. The user would "tune" to the desired frequency slot.

If lower-quality, nonstereophonic sound is used, one television channel could carry 600 sound channels simultaneously. These might be used for continuous news broadcasts, weather forecasts, a channel repeatedly telling the time, continuous sports reports, stock market reports, community information reports such as theater and movie listings, and shopping news. More than 500 sound channels would still be left for radio plays, talks, poetry, language teaching, and other programs, all of which might be repeated continuously for one day and then replaced by new programs.

Because sound occupies so little channel capacity compared with the television image, a television program could be transmitted with more than one sound track. This would make it possible to transmit a program dubbed in several languages. The user of a suitably equipped set would press a language button to select a particular language. This could provide a service to foreign-speaking residents. It gives the attractive possibility of relaying high-quality foreign television programs in several languages.

An American television channel delivers thirty television images (frames) every second. (Television in many other countries operates at 25 frames per second.) Just as you can inspect the individual frames of a movie, so it is possible to inspect a still television picture. To do this the television set must have electronics that can select and hold an individual frame. The equipment that does this is called a *frame grabber.* Television sets with frame grabbers have been marketed. They are sometimes used by sports enthusiasts for freezing a moment in a game to study the players. Frame grabbers are rapidly dropping in cost. It is possible to broadcast television in which each frame is *an entirely different picture.* For example, one frame might contain a weather map, a news picture, a page of a news "magazine," an advertisement, a picture of community events, or whatever the audience would find useful. The user could employ a small keyboard to request the picture he wants to examine. If the average time taken to respond to this request is five seconds, one television channel could carry 300 such pictures continuously. They could be changed frequently at the television studio.

Instead of television pictures, data could be transmitted so that the user could select displays of letters and figures. These might give temperature, barometer and time readings, weather forecasts, news summaries, stock market summaries, the ticker tape, movie listings, and so on. One television channel could carry many thousands of separate frames of information.

As with the other possible functions of cable television, it is far from certain that the full capacity will be utilized. It is far from being utilized today. Moreover, it is uncertain who would pay for such a proliferation of programs, although the programs need not be expensive. There could be frequent reruns of old programs, as well as music without broadcasters. Foreign programs could be relayed via satellite.

Nonetheless, it is worthwhile calculating how the capacity *could* be used.

MUSIC LIBRARY

One appealing consumer service would be a cable television music library.

Suppose a music "library" that plays an extremely large number of pieces of music continuously is set up. A tape deck, like a videotape machine, might be used to play 100 pieces of stereophonic high-fidelity music continuously. Forty such machines might be used with different playing times to cater to the different lengths of the works; this would mean that 4000 music tracks were playing simultaneously. The music could include every major symphony, opera, and concerto, and a vast amount of jazz, pop, and show music. This collection of music could be transmitted continuously to all the cable television systems in a city. It could be relayed by satellite to television systems throughout a continent.

The majority of sound channels on a television cable could carry music determined by the programming authorities. The majority of subscribers would be content most of the time to listen to what was offered, especially if a large number of channels existed, as indicated earlier. At any one time, a minority of subscribers would be playing a symphony or other piece of music *requested* from the library.

Suppose there are 3000 subscribers on a cable and not more than half have their sets switched on in the peak hour. Of these subscribers, not more than 20 percent are likely to be listening to sound-only programs, and of those that are, only 25 percent are listening to *requested* music; the rest are listening to the preselected channels. Thus the needs of the subscribership could be met by the seventy-five request channels and twenty-five preselected channels of music on *one* television channel.

If there were a million music subscribers on a continent, such a service could make a handsome profit. The Musak Corporation plans to distribute several channels of music via satellite.[2]

It seems possible that advertising might be added at the end of a piece of music to fill the gap between the end of the music and the end of the tape. The advertising might be designed to appeal to music lovers or high-fidelity enthusiasts, who may constitute a highly selective market for certain products.

SELECTION OF OTHER PROGRAMMING

The idea of being able to request one's programs instead of merely accepting broadcasts, which are not usually planned for the most selective audiences, is appealing. It is economically feasible to request sound-only programs from a large assortment. With movies the choice would be limited, because the television cable cannot carry a large number of movies. If higher-capacity (optical) cables were used or if one cable served a smaller number of homes, the choice could be greater. Eventually today's *telephone* cables and exchanges will be replaced by optical cables, and switching facilities will make it possible to dial video sources rather than mere telephones.

PAY TELEVISION

While even forty-channel cables do not have enough capacity to allow their many subscribers to select movies freely, they could permit a limited degree of selection. In one proposed system subscribers select programs from a guide and inform the system what they want to see at a given time. Because the selections available are chosen so that most subscribers will pick certain items (for example, currently popular movies and sports events), most requests can be met. The most popular requests are determined by a minicomputer. Some subscribers must be disappointed, but they are told that their selection can be scheduled at another time and are offered a second choice.

This scheme is one variant of a much discussed cable service, Pay Television. The cable television industry is searching for services that would increase cable revenue and Pay Television is usually at the top of the list. Pay Television, also called Pay Cable, Subscription Television, and euphemistically, Premium Television, would attempt to provide exceptionally desirable programs and charge specifically for them. Such programs would include sports events and first-run full-length feature movies without advertising. Home Box Office is a cable television service that has been highly successful in the United States, providing movies and sports events for a fixed monthly fee. Home Box Office programs are now being distributed to cable systems in many parts of the United States via satellite.

TALKING BACK TO YOUR TELEVISION

In 1972 the government ruled that U.S. cable television systems should be constructed with reverse-channel capability. This makes possible systems in which the users could talk back to their television sets. To do so one needs an inexpensive keyboard that attaches to the set. A pushbutton telephone keyboard would suffice for many applications. So would devices like those now sold for playing games on a television set.

This creates an interactive medium like Viewdata but with one major improvement. Viewdata can display only text, figures, or crude diagrams composed with blocks. Cable television can display images like color photographs—any still television frame.

One's daily newspaper in the future may occupy one such television channel, perhaps with color photographs and news items assembled in many countries. The users would *select* the news that interested them. Interactive advertising could be used, as well as shopping from the home, color catalogues, computer assisted instruction with color pictures, and so on.

The cable television industry refers to such uses of the cable as *subscriber response services.*

The Warner Brothers' Qube cable system in Columbus, Ohio, allows viewers to respond to television programs with five response buttons on a calculator-sized console. The screen tells the viewer the meaning of the five buttons when they are to be used. For example, one person showed football games which were interrupted to ask the viewer questions: If the viewer were the coach would he or she choose, say, a quarterback option, a fullback hand-off, or a sprint-out pass? A computer shows a summary of the viewers' responses on the screen as they respond. The program then shows the play the coach followed.

Viewers also punch their verdicts on amateur talent shows. One showed a group of youths called Bo's Burping Band who belched tunes accompanied by a Jew's harp. The computer summarizing the audience response decided that the band be yanked off in mid-burp. *Newsweek* suggested that the same technique should be applied to guests of the Johnny Carson show.

A similar system could be used for bidding for merchandise on televised auctions, responding to politicians, participating in all manner of game shows, registering opinions on local community issues, and so on.

DIALOGUES WITH PICTURES

Many appealing forms of dialogue can be imagined in which the responses come in the form of a still color picture on the television screen rather than in the form of a digital response (with letters and digits). For example, suppose a woman wants to continue planning her vacation, or perhaps she is dreaming about a vacation she is unlikely to have. She uses her set to contact the local travel-agent computer, having been instructed how to do so by advertisements on other channels. The travel-agent computer has her account record, which states how previous dialogues have concluded. It displays the following message on her screen.

```
GOOD EVENING MRS. SMITH.
WOULD YOU LIKE A CONTINUED PRESENTATION
ON ONE OF YOUR PRIME SELECTIONS:
1. PERU?
2. BOLIVIA?
3. HAWAII?
4. EASTER ISLAND?
IF SO PLEASE KEY THE ABOVE NUMBER
```

She presses the "4" key on her keyboard. The "4" travels over the digital input channel, via the cable computer, to the travel agent computer. The travel agent computer responds,

```
EASTER ISLAND.
WHICH OF THE FOLLOWING WOULD YOU PREFER:
1. SCENERY?
2  ENTERTAINMENTS?
3. HOTEL INFORMATION?
4. SHOPPING?
5. TIMETABLE?
6. TO MAKE RESERVATIONS?
PLEASE KEY THE ABOVE NUMBER
```

She presses the "1" key, and the screen says,

SCENERY.
PLEASE PRESS THE # KEY TO CHANGE SLIDES
AND BEGIN PRESENTATION.

She presses the # key. There is a pause of about six seconds, and then a magnificent aerial view of Easter Island appears on the screen. The women looks at color pictures for the next half hour and then presses a key labeled END. A message appears on the screen, trying to sell her a booking. She responds negatively and switches the set off.

All manner of applications for dialogue with pictures can be imagined. It could be particularly valuable for teaching, especially with children. Madison Avenue might be a major source of material for such a channel, but other, longer-lasting "programs" would be steadily amassed, as books are in a library.

HOW TO MAXIMIZE VALUE

To make this new medium grow in social value as rapidly as possible, the provision of source material should be made open and entrepreneurial. Any individual or corporation should be able to be an information provider, or program provider, and when the provider's material is used, royalties should be paid. The growing army of computer hobbyists should be encouraged to become contributors and make money. At the same time the maximum incentive should be given to publishing companies, encyclopedia companies, consultant firms, retail chains, horticulturists, real estate firms, and so on, to use the medium.

Like cable television itself the medium should contain a mixture of local community information and nationwide or international information. It should serve the local tradesmen, real estate brokers, employment agencies, and politicians, and also be connected to switched networks that give access to nationwide information resources, as with the British Viewdata scheme.

A MORE HUMANIZED MEDIUM

The new television can be a more humanized and friendlier medium than conventional television. A person arriving home can obtain the news summary or sports results *when convenient,* not at some fixed time. The set responds to requests. There should be many information and program sources so there will be something to delight and match the needs of the user.

Because it can have many channels, some of them local, cable television encourages a different relationship between the broadcaster and viewer. Instead of an audience numbering tens of millions and programs dominated by ratings, many of the cable channels will carry programs for a small group of more definite personality. The broadcaster will be concerned with that personality. When a large number of channels are available, enthusiasts can create programs for enthusiasts, scientists for scientists, women's groups for women's groups, hobbyists for hobbyists, poets for poets, and so on. The specialist programs may be far less expensive than the glossy mass-audience programs. The medium becomes more like magazine publishing. The cable companies are concerned with how many subscribers will pay the extra money for the cable when they already have through-the-air television. Many cable subscribers will want the set for their own specialist programs.

Instead of the big commercial networks' desperate search for the lowest common denominator in the audience, television with many channels encourages a different kind of search. How can the operator fill the channels so that many different types of people find something of more value to them than conventional television?

Executives of mass-audience network television ignore their own taste as a matter of course. Producers for specialized audiences are using their own personalities to communicate. *Newsweek*'s Douglas Davis describes it as "closer to the kinship that exists between friends, or between essayist and reader."

Technology, then, can bring diversity, and diversity changes the culture nature of the medium.

Chapter Six

NEWS

The medium modifies the message so we must have a wide diversity of media.

New communications technology can engender new types of public news media. The introduction of new forms of news dissemination has often brought dramatic upheavals in the structure of society and its institutions. New electronic media will bring further upheavals, but these may be changes for the better.

When power printing presses were coming into use in Europe and newspapers were beginning, Napoleon commented: "Cannon killed the feudal system; ink will kill the modern social organization." In the 1840s telegraph channels spread across Europe and new printing machinery caused newspapers to fall dramatically in cost until many households could afford them. In 1848 there was an astonishing series of rebellions and revolutions that shook almost every regime in Europe and brought an unparalleled degree of social change. There were multiple causes, related to the spread of the industrial revolution and urbanization, but the attitudes of the time were greatly affected by the new communications channels, which bypassed the traditional leaders. Emotional ideas, such as democracy and doubt about the divine right of kings, swept across Europe. Individuals suddenly perceived their roles in society differently and became prepared to agitate and fight for their new perceptions.

In the late 1920s and the 1930s radio became widespread in Europe. Hitler realized its power for mass communication and used it almost without opposition, causing a civilized people to behave in a fashion nobody would have thought possible.

The sudden introduction of transistor radios to developing countries and illiterate people often caused unprecedented and unpredicted upheavals. A major effect in some countries was to cause people to move from their well integrated and relatively happy villages to the towns where they now imagined there must be excitement and opportunity. The towns swelled with the influx of untrained people, who found it difficult to obtain jobs, food, or accomodation. Slum areas grew up with appalling squalor and human misery.

The 1960s were an extraordinary time in the United States. There has probably never been a decade of such political and social iconoclasm. Every institution in society was challenged, often with riots, campus disorders, and street violence. There was probably a greater degree of social innovation than at any other time, resulting in new roles for women and minorities, new controls on industry, numerous social programs, and ecological controls intended to bring technology into greater harmony with nature. Some social commentators have said that the 1960s saw a greater rate of social change than any other decade in human history. Again, the reasons were multifold, but the 1960s were the first decade in which the majority of Americans owned receivers for what has been described as the world's most powerful communications medium: television.

President Charles de Gaulle of France commented on his political opposition early in the 1960s: "They have the press. I have the RTF, and I intend to keep it." The RTF is the state broadcasting system, office of Radiodiffusion-Télévision Française. De Gaulle controlled the television and used it with great skill. "Français, aidez-moi . . ." De Gaulle's was the first long-lived and stable regime in France for decades.

Stalin in the 1950s said, "The cinema is the greatest means of mass agitation. The task is to take it into our hands. If I could control the medium of motion pictures I would need nothing else in order to convert the entire world to communism in a very few years." Khrushchev made similar comments about television in the 1960s.

The media and the news that people receive do much to determine the social and political values of society.

The average American eighteen-year-old has spent 12,000 hours at school but 15,000 hours watching television. Programs are designed without regard to the awesome responsibility represented by this figure; they are designed to maximize ratings, which in turn maximizes the revenue derived from advertising. A peak-hour program with an audience of 5 million is considered a failure. To achieve the largest audience, producers must employ all manner of sensationalism. In the 1960s it was found that audiences were attracted by antiestablishment films. So the films of the era attacked the government, the police, patriotism, the army, and the intelligence agencies, portrayed industry as crooked, and homosexuality as normal. Conventional marriage was too dull for the screens, so society watched all manner of deviations from it. Cinema portrayed the wildest of sexual inventions. The Vietnam war was unpleasant, and television could maximize its ratings by ridiculing it. Rightly so, you might say, but no other war in history has been fought against such a background. The home screens even showed propaganda films made by the North Vietnamese for American television. Senator Dole asked: "Does freedom of the press include the right to incite mutiny?"

The inversion of traditional values was done neither maliciously nor philosophically. It was done by businessmen and technicians responding to the technology and the need to maximize ratings. It was a byproduct of brilliant inventions in engineering and the corporate structures that employed them.

If we have different values and hence live differently in the future, one reason will be that the nature of our information media has changed.

TECHNOLOGY CHANGES THE NEWS

The technology of news dissemination affects both the content and the form of the news.

For 350 years after Gutenberg invented the printing press, news was disseminated largely by town criers and word of mouth. The Gutenberg machine was a converted wine press, and was too slow to create daily newspapers.

In the eighteenth century the printing press was used to print items for community bulletin boards and a few pamphlets that were the forerunners of today's newspapers. Early journalists like Daniel Defoe, Jonathan Swift, and Tom Paine created pamphlets that colorfully advocated their personal causes. To journalists of their era, the suggestion that they should report news in a balanced and objective fashion would have seemed ridiculous. The printing process was too slow and expensive for any such extravagance.

In 1810, for the first time, a steam engine was used to

operate a printing press; it could produce 200 printed sheets an hour. Soon afterward the rotary press came into existence. The first rotary press was smashed by angry crowds, but the *Times* of London installed one that could print 2400 pages an hour. The rotary press eventually lowered the cost of newspapers until they were no longer an elitist commodity but one many householders could afford. Weekly newspapers became common, and then dailies. In the early 1830s city newspapers cost sixpence. A one-year subscription cost the average householder more than two weeks' pay. The cost dropped steadily until many householders received daily information for the first time.

By 1900 there were 16,000 newspapers in the United States. They received news items from distant sources by telegraph, but most of them served small local communities, because transportation was not fast enough or frequent enough for wide distribution. Subscribers were easily enticed by sensational news, and as other news media did not yet exist, readers had no way of checking their newspaper's accuracy. This was an era of legendary journalists, like Damon Runyon, famous for the stories they invented, and multimillionaire newspaper owners, like William Randolph Hearst, who made their fortunes from melodramatic fiction presented as news, sometimes for political causes.

Presenting fiction as news is unlikely to succeed if there are multiple sources of information and fast, nationwide communications. In time technology brought an end to the era of ingenious invention of news. The airplane made it possible for national dailies and weeklies to compete with local papers, and radio gave an alternative source of information. The recipient could compare news from different sources. Thus, only in the twentieth century has the responsible, objective journalism we expect today flourished. It would not be reasonable to say today what Thomas Jefferson wrote in 1807: "The man who never looks into a newspaper is better informed than he who reads them, inasmuch as he who knows nothing is nearer to truth than he whose mind is filled with falsehood."

Radio brought *instant* news. It did not require the lengthy time for production and distribution that newspapers needed. Newspapers regarded radio as a threat, and in the United States they took political action that severely limited the length of early news broadcasts. Broadcast news was also inhibited by the fact that radio advertisers preferred entertainment to news. However, radio outdid the newspapers in giving weather reports and racing results and in the initial announcement of sensational events. Melodramatic headlines and newsboys' cries in the streets became less effective in selling papers. Newspapers had advantages in their ability to give details too lengthy for radio and in their diversity, which enabled readers to scan many news items quickly, deciding which to read.

As radio technology improved, its role in society changed. In the early 1930s only 40 percent of American families owned radios, whereas almost 90 percent subscribed to daily newspapers. Radios dropped in cost until by 1950 virtually all families had one, but fewer received daily papers. Later, car radios became common and stations broadcast traffic reports, frequent time checks for harassed commuters, and only the briefest of news items. A Rand survey of radio stations in 1969 revealed that the *longest* news item of the average AM station was only 240 words.[1]

The transistor and its successor, the integrated circuit, made it possible to manufacture inexpensive radios. These were cheap enough to be sold in vast quantities to the people of developing nations who could not read. A radio without a tuning unit is extremely inexpensive, and the governments of some developing nations distributed such radios to people who had no newspapers or television. Nontunable radios had the advantage to some governments that people could pick up transmission only from the one government station.

In the 1950s television spread with remarkable rapidity. With this medium the news was more vivid and emotional. It could show human emotion and human misery; it was good for action and for visual news such as fires and plane crashes. It had great appeal to a certain cross section of the public who found it easier to watch television than read. Marshall McLuhan expressed the difference in content between printed information and television with his famous exaggeration, "The medium is the message."

At first technology permitted only a few television channels. In some countries they were commercial channels competing intensely with one another. The news broadcasts, like any other programs, had to sell products and make money for the advertisers. The god of commercial television is the *rating:* How many people are watching the program?

Broadcasters and the "media consultants" who advise them learned how to maximize the ratings of news broadcasts, and again the medium modified the message. The definition of news changed from "What people *should* know" to "What people *want* to know." The general public are most likely to watch the news if it includes crime, sex, gimmicks, and "happy talk"—news that is long on fun, sensation, and trivia but short on analysis and depth. They want plenty of fast-moving stories with high visual appeal. No item should be longer than ninety seconds; sixty seconds is considered ideal. There should be eye-catching graphics and breezy conversational reporting. Complex nonvisual stories are cut to be very brief or omitted. On highly competitive commercial television, news is show business.

There is a major difference between the news commentary and current affairs programs on American and Latin American commercial television and on noncommercial networks such as the two BBC channels in Britain and the two NHK channels in Japan.

Just as past technology changed the news and the news changed society, future technology will bring its own changes. There are several technological advances that will have a major effect.

MANY CHANNELS

First, it is now possible to provide a large number of television channels. Some cable television systems have forty channels. Full use of the UHF (ultrahigh frequency) radio frequencies allocated to television in the United States could provide more than forty channels. A communications satellite weighing several tons, designed to fit the cargo bay of the space shuttle, could broadcast one hundred channels directly into the home. If as much money were spent on such satellites as on television stations in the United States today, the satellites could broadcast *hundreds* of channels to home receivers. The most imminent increase in the number of channels will probably come from cable systems interconnected via satellite. There are already many earth stations for this purpose with large dish-shaped antennas receiving television relayed via satellite.

If forty television channels go into the home, it is unlikely that all of them will be commercial. It will be difficult at first to find enough material to fill forty channels, but the possibilities are exciting. First, there are excellent news and current affairs broadcasts made in foreign countries, and satellites can transmit worldwide. Programs on world affairs should be made available throughout the world. Broadcasts from other parts of the world would give the public entirely new perspectives. World news is reported very differently in different countries, and the commentary on it varies greatly. (For example, I crossed the Atlantic Ocean frequently at times during the Vietnam war, and the difference in reporting between Great Britain and the United States sometimes made it seem to be two different wars.) If a program like the BBC's *Panorama* had a world market, a larger budget would be available for its production and research and its coverage could be correspondingly expanded. Worldwide distribution of current affairs television would do much to forge a better understanding between peoples of the world.

At the same time as becoming *international* via satellites, television can become *local* via cables. Local town meetings, local sports, school board meetings, plays, and discussions are being broadcast on some cable television systems now. Television serving the local community makes much sense: It can help stimulate interest in local affairs and government and build a community spirit; critical issues such as zoning can be explained and debated on the air by local personalities.

Television sets, then, should receive not only national television, as they do today, but also broadcasts from other countries and local community programming. Only some of the channels need be commercial, and only some of the news hyped-up, competitive, show business news.

It has often been suggested that proceedings of governments should be televised. It would cost very little to have fixed cameras operating in Congress, the Houses of Parliament, and the meeting places of local government. Occasionally this has been done, and the result is usually boring to most people. However, if it were normal practice whenever government meets, some politicians would be more responsive to their constituents, knowing that at least a small number of the constituents would be watching.

When television has a large number of channels, such procedures are practical. It may be better to split some of the television capacity into hundreds of sound channels and broadcast meetings by sound only. The most important parts of a day's proceedings could be broadcast on several sound channels with time differences, so that a listener at a given time would have a good chance of finding what he wants to hear.

If there are many channels, there should be a diversity of different types of news reporting and commentary, with different styles and flavors aimed at different types of viewer. In England there are nine national daily newspapers, which differ greatly in their style. One wit summarizes the differences as follows:

The *Times* is read by the people who run the country.
The *Mirror* is read by the people who think they run the country.
The *Guardian* is read by the people who think they ought to run the country.
The *Morning Star* is read by the people who think the country ought to be run by another country.
The *Daily Mail* is read by the wives of the people who run the country.
The *Financial Times* is read by the people who own the country.
The *Daily Express* is read by the people who think the country ought to be run as it used to be.
The *Daily Telegraph* is read by the people who still think it is.
The *Sun* is read by the people who don't care who runs the . . . country as long as she's got big tits.

With many television channels, the same diversity of news reporting ought to be expected from television as well.

THE GATEKEEPERS

Not all news can be made available to the public; there is too much of it. Most news film footage is thrown away from a cutting room floor. On a large newspaper, 85 percent of all of the stories reporters write are never used. Someone, usually with the title "news editor" or "wire editor," has to decide which news is used.

Social scientists refer to the various individuals who select news stories as "gatekeepers." Executive editors and publishers are known to the public, and their power is respected. Reporters may have their names on stories and are thought of as doing glamorous work. The gatekeeper is usually unknown to the public, but has remarkable power. Besides deciding which routine stories will be used, the gatekeeper sets a pattern that tells reporters which stories it is pointless for them to report. Selection of news depends to some extent on the gatekeeper's personal views of the news or its audience. It also depends on the mechanics of production: The newspaper printing system has to be fed slowly and steadily throughout the day. A Rand study of wire copy showed that the most important single factor determining whether a news item is used is its time of arrival. It is more likely to be used if it arrives early in the day. Another bottleneck occurs in the teletype services that carry reporters' copy. The Associated Press and the other wire services carry the copy of many reporters to many newspapers. The lines typically operate at forty-five words per minute. If many reporters are sending much copy from a given area, the copy can become severely delayed because of the low transmission speed.

Thus, mechanical factors such as printing speed, page size, and teletype transmission rates affect the content of newspapers. By permitting faster printing equipment, computerized editing, and fast data transmission, new technology can remove restrictions other than the size of the newspaper itself. Size restrictions can cause severe selection problems, especially on weekly papers and magazines.

COMPUTERIZED NEWS

The Bible, which took Gutenberg five years to set into type, could be transmitted in one half second on one satellite channel. One large computer storage unit can hold the contents of a vast library with ten miles of shelves. The combination of electronic storage and transmission makes possible a new way of handling news stories.

Reporters could write what they wished, regardless of teletype speeds or gatekeepers. Their reports would be transmitted to computers, where they would be filed and indexed. In composing a newspaper, an editor would work visually from these files, the text being edited on a screen. The results could go directly into the printing process by means of computerized photocomposition. This would be a major step in automating and speeding up the production of newspapers. Most reporters' copy would not be thrown away if not used. The files of copy would remain available for composing weekly news magazines or for research at a later time.

The New York Times has an information retrieval system that makes it possible to search for past news items using a computer screen. The user might ask, for example, Has the Club of Rome made any statements about nuclear fusion being a future solution to the energy crisis? The computer searches all past copies of the *Times* back to a given date. To make this possible, abstracts of all the news stories have been previously created, and the computer has built elaborate indexes that make possible a rapid search of all past editions. The system stores not only *New York Times* reports but also those of other important newspapers and magazines.

Users with screen units thousands of miles away have access to the system via telecommunications. In the future there will be many such systems, accessible throughout the world via data networks. Both the users and the newspapers searched could be anywhere in the world.

The New York Times system stores reports that *have been printed* in newspapers—i.e., after the gatekeeper has eliminated 85 percent of the stories. In many ways it would be easier to store the reporters' original unedited copy. A reporter wanting to place copy in the system would type a brief abstract for it, and it would then automatically be filed and indexed for future use. A vast amount of computer storage would be needed, so to lessen the requirements old reports might be in microfilm libraries. However, computer file technology is improving and decreasing in cost at a very rapid rate.

DO-IT-YOURSELF NEWSPAPERS

Using the television channels, prodigious quantities of print could be delivered by telecommunications to the home. A television channel has 100,000 times the transmission capacity of a teleprinter. The consumer would not want the massive quantities of print that could be delivered. Instead, each consumer's equipment could allow selection of what he or she *does* want: stock market data, weather reports, news headlines, or much more detailed information such as the minutes of a town meeting, the eyewitness reports of a plane crash, the congressional debate on tax reform, the statistics on farm production, or the latest government bribery scandal.

Under such a system news would no longer be limited. Without gatekeepers the passage of news from reporter to consumer would be untouched by human bias, and a consumer who wishes to can have access to the copy of different reporters.

Of course, only a few consumers would want news this raw; most of them would want it processed and preselected. Thus, editors would perform a changing but increasingly valuable function as technology made possible *multiple* news media with different types and different levels of editing.

When data networks are accessible from the home, the consumer will be able to ask for any news, as users of *The New*

York Times information system now can. If he or she wishes, the consumer can command a computer to search for items of interest.

Another possible approach is for the user to leave with the computer a list of types of news he or she wants or subjects of interest. The consumer can register a profile of his or her interests, and if there is a home printer, it will automatically type the news that was requested.

News delivered electronically to printers in the home may sound expensive. In fact, it could cost less than today's newspaper distribution. The expenses of a large newspaper are $70 million per year, two thirds of which are for newsprint, production, and distribution. In the United States a total of about $5 billion per year is spent on newsprint, production, and distribution, but a satellite system for direct broadcasting of data to homes could cost less than $200 million. Even in the absence of such a satellite, sending the data by telephone or television cable would cost a fraction of today's physical distribution costs. The home receiver could either display the print on the television screen, allowing the user to select the "pages" of interest, or else the user could have a small printing or copying (i.e. "facsimile") machine. If the latter were a mass-produced consumer item, it could be manufactured for about the cost of a color television set.

In 1976 the BBC started an experimental service, called Ceefax, for broadcasting "magazines" of data, which are displayed in six colors on the home television. France has a similar service, called Antioche. The viewer using these services now has access only to a small number of pages, but the services could be designed so that several complete television channels are used and many thousands of pages are broadcast. The service displays charts, such as weather forecast maps, as well as text and tables. A viewer can display these at any time. A conventional television set can be used with a small adapter.

MULTIPLE MEDIA

In summary, technology can now give us multiple news sources. We are not suggesting that computers will replace newspapers or that satellites and cables will destroy commercial television or radio. All of these have a role to play in a richly textured fabric of information sources. The medium modifies the message, so we must have many different media.

Some television news may come via the "hot" medium of large-screen commercial television, expensively produced and edited, highly competitive, eye-catching, sensational, broadcast on far-flung networks. Other news will come via the "cool" medium of small-screen black and white channels relaying government meetings or debates. Some will be broadcast headlines, text, and data, viewable in color at any time the set is switched on. Some news channels will be national, as today; others will be local, giving town and community information; still others will be international, showing how different nations view world affairs and perceive each other.

News will be summarized and predigested for people short of time. But complete news will also be available—the original reporters' copy, indexed, computer-retrievable.

As in the past, the new news media will change society. None, perhaps, would make a change greater than that described in the next chapter.

Chapter Seven

PUBLIC RESPONSE SYSTEMS

Technology can now give us a way to talk back.

A major cause of unrest in society is people feeling that they do not have a say in how things are run. The government seems vast, bureaucratic, and unresponsive to the needs or wishes of individuals. Voting once every few years seems to bring about few important changes. There is no effective way to talk back to the politicians.

Technology can now give us a way to talk back.

Imagine a small box with keys on it, cheap enough to be in every home. It transmits a simple message to a distant computer. The message can say merely *yes* or *no*, or it can be a single digit. The twelve keys of a pushbutton telephone could be adequate for this purpose.

The machine is used in conjunction with television programs. After a political speech or during a current affairs or news program, the public is asked to respond. One may be asked to say *yes* or *no,* or to rate one's opinion of something on a scale from 1 to 7, by pressing a key. A small computer in each area scans the results, accepting only one signal from each such device. A light appears on the device, telling the user that his or her message has been received correctly. The area computers count the various responses and transmit the results to a central computer. Within minutes the television displays the results: For example, 63 percent of the viewers with devices responded; 22 percent voted *yes;* 78 percent voted *no.* A news commentator could obtain an immediate reaction to a President's speech. The President could ask for a public response in midspeech. A town authority could ask for the local community's reaction to a zoning proposal. Current affairs programs could explore shades of public opinion. Imagine a local politician being interviewed on television and the interviewer saying "I'm not sure that I believe you, Mr. Mayor. Let's see what the audience feels."

The devices could also be used for entertainment: Say who you think murdered the fat police chief by setting him in a lucite block and mailing it to his mistress.

Television could broadcast the proceedings of government live, with audience responses. The responses might be displayed in the House or Senate. Some politicians would be likely to play to the gallery. They might present their arguments more clearly and persuasively. The gallery would be there, sometimes millions of people, and they would respond. This may not always improve the debate but there would at least be *communication* between politicians and people.

For the responses to be statistically valid, it is not necessary for a high proportion of the public to have transmitting devices. Public opinion polls are usually based on a few thousand representative samples. U.S. television ratings, on which billions of dollars of advertising expenditure is based, are drawn from the sampling of 1200 carefully selected television sets. Thus, audience-response television could begin in a relatively inexpensive way, with a few thousand devices. To give the public a feeling of participation, there would eventually be large numbers of devices, cheap enough so that the general public can afford them. The sampling would become seriously biased if only rich people had the devices or if only persons with certain attitudes used them.

Audience-response television is a powerful form of communication that could become part of the democratic process—more so than public opinion polls today. It has been suggested that government referenda be taken using such devices, but such referenda would be unlikely to have the effect of law until almost 100 percent of the public had suitable machines and the system was designed with adequate security. It is more likely that audience response devices will be used not for elections or referenda but for continual testing of public attitudes.

MECHANISMS

A variety of different mechanisms could be used for audience response. The best choice of mechanism will change as technology changes.

The device might look like the keyboard of a pushbutton telephone with the * and # keys labeled YES and NO. The machine might emit a distinctive tone to indicate that a response has been received. In this case the device could be a conventional telephone handset. It could be the keyboard of the British Post Office *Viewdata* television set. The receiving equipment should be designed so that only one response can be received on one telephone line during a given period.

Alternatively, cable television might be used, with the response box connected to the cable. This is done in the Warner Brothers Qube system operating in Columbus, Ohio.

A third form of mechanism, independent of either the cable television or pushbutton telephone, is packet radio. When used, the device transmits an instantaneous burst of radio signal, which carries the coded response. The burst carries the identification of the device, so that a receiving computer will not accept two bursts from the same device in a given period.

Any of these home devices could be inexpensive if mass-produced. With a combination of the three techniques, audience-response capability could be given to an entire country. It would cost less to equip every household in the United States with an audience-response device than it cost to land the Mariner spacecraft on Mars. It would cost less to equip Canada than the Olympic Games cost in 1976. It would cost less to equip my hometown than public swimming facilities cost.

OTHER USES

Audience-response devices could have many uses other than opinion surveys. We discuss some of the possibilities in Chapters 5 and 14. One class of applications has a high potential revenue that would pay many times over for the introduction of the simple devices we have described. When stores, mail order firms, and sales departments advertise goods on television, they could request that interested parties make a response on their machine. The computers would be programmed to identify the owner of the responding machine and relay the name and address to the advertising party. The advertiser could then telephone, send brochures, or send goods and charge them.

Thus, when used in conjunction with still-frame television, audience response devices could permit home shopping. The total expenditure on advertising in the United

States is over $20 billion per year. Cable television companies might attempt to obtain part of this revenue by offering audience-response facilities.

EFFECTS

There is little doubt that audience-response television would have a major effect on the body politic. Authorities argue about whether the effect would be entirely good.

We have passed through periods of social violence, campus chaos, and tempers exploding into riots in the cities. Audience-response television might make such outbursts less likely by giving people a nonviolent outlet for their discontent and by giving authorities a means of measuring it.

Both engineers and sociologists are familiar with *feedback* mechanisms for controlling complex systems. To maintain stability in social or other systems it is necessary to have good measurements of those forces that cause instability. The controlling mechanisms or government can respond to the measurements when they indicate undesirable deviations from a stable state. Some authorities claim that audience response television could act as a form of social instrumentation, measuring the tensions and thus helping to maintain stability.

We have seen spectacular development of mass communications, but it is all one-way communications, from a central authority to the public. The public cannot reply. In Victorian England concerned citizens regularly wrote letters to the *Times.* But it does no good to write a letter to the television authorities. Society has been defined as "people in communication." If society seems to be disintegrating, it is probably because the communications channels are not adequate. In a small village people are truly in communication: They meet in the streets, pass on rumors, and can have plenty to say in village affairs. Modern communications have given us national horizons, not village horizons. News comes at us from far-away faces on television. We pay massive and growing taxes to distant governments. Marshall McLuhan talks of a global village, but the global village does not really work, because there is only one-way communication.

The flowering of democracy—government of the people, by the people, for the people—took place in Athens and the Greek city-states. The units governed were usually not more than ten thousand people. All citizens could attend the meetings held in the city square and express their views. The voice of the poor man was heard along with the voice of the rich (although women and slaves were excluded). When voting on issues took place after such debates, all persons voting were well informed. Greek democracy worked because men had sufficient leisure time to meet, debate, and become well informed. They had the leisure only because of the labors of women and slaves. Today we are entering an age of automation when women should be equal to men and only machines should be slaves. People will have enough leisure time to become well informed, and they may have difficulty filling that time.

Prior to the formation of modern democracies, the major theorist of democracy was Rousseau, who wrote, "No law is legitimate unless it is an expression of the general will, a consensus of the whole community. No man can enjoy full moral responsibility, and so be really a man, unless he participates in the formation of the consensus by which he is legally bound. This means that he must assemble with his fellow citizens at periodic intervals, and personally vote on each and every act of legislation." Authorities of our age have claimed that Rousseau's view, while theoretically desirable, is not practical today. It is appropriate for Athenian city-states and perhaps for New England town meeting, but not for governing millions of people. Public response telecommunications, however, could make it practical. The use of such systems might one day be regarded as essential to a satellite-age democracy.

USER IDENTIFICATION

The system we have described does not identify the individual who responds nor the individual's household. It is able merely to summarize the responses statistically. With computers a more complex system, in which the households are identified, would be possible. The responses could then be used to establish a census profile. The system could relate attitudes to social groups, income groups, areas, and so on, thus giving a much more comprehensive understanding of types of public reaction.

A particularly interesting form of response might relate to how people would like to see their taxes spent. It is possible that today's expenditure patterns would far from match the public wishes. Views on tax expenditure could be related to the amount of tax *paid* by a household, and statistics could be compiled on how individuals want *their own* tax payments spent, large or small.

There are many possibilities for uses of public-response systems.

BADLY INFORMED RESPONSES

There is some opposition to the idea of giving opinions from the home on public issues, on the grounds that the majority of the public are ignorant about these issues. Some are opposed to audience-response television because ill-informed responses might unduly influence the political process. This is a valid cause for concern, and the answer lies in the responsible use of the media. The media should ensure that the public know the facts and the arguments before voting. The debates prior to voting in the Greek city-states sometimes went on for days. It could be disastrous to have such audience responses coupled to today's commercial television, with its "news for people who don't like news." A sensible rule might be to permit audience responses on government issues only at the end of a program at least one uninterrupted hour in length explaining those issues, and to require that substantial noncommercial channel time be used to equip the public with the information necessary for fulfilling their responsibility.

Another concern is that persons will always vote for their own short-term selfish interests. They might vote for low defense budgets, no tax on gasoline, and free beer. They might vote for decisions that would lower their taxes but harm the country. If the average income is $8000, the mass vote might be for very high taxes on income over $12,000. Some authorities cite decisions that are advantageous in the short term but disadvantageous in the long term. Probably this question also depends upon the public being well enough informed about the overall effects and benefits of today's decisions. If they are well enough informed, voting for their own overall long-term interests will probably benefit the country. Television is the ideal medium for making them well informed, though not as it is being used today.

Another concern is overemotional reaction. The public can react suddenly, and with little thought, to an emotional issue. This is particularly likely to occur with issues relating to foreign aggression. Politicians would learn how to evoke emotional reactions, just as they learn how to manipulate crowds. To avoid one-sided emotion it is desirable that multiple programs, reflecting different political opinions, should be seen and responded to. In other words, there should be debate, not merely a single electronic response to a demagogue. An important requirement of feedback control mechanisms is that they do not react instantly to responses that are likely to fluctuate in value. Instead there is a built-in time delay; the responses are averaged over a certain period of time. The mechanism is *damped* to ensure smooth running and avoid wild fluctuations. Wild fluctuations do occur in public reaction to emotional issues. It would be highly desirable not to go to war or to overreact on the basis of them. There are even some situations in which national leaders need to be left alone to take powerful action unhampered by public debate. Churchill commented in 1941 that "Nothing is more dangerous in wartime than to live in the temperamental atmosphere of a Gallup Poll, always feeling one's pulse and taking one's temperature." While this is true in war, it is debatable under what circumstances it is true in peace. Henry Kissinger, speaking about absence of leadership in foreign policy in 1974, commented that a tree cannot grow if you pull it up to examine its roots on television every night.

The real issue here is the viability of democracy. Dwight D. Eisenhower defined democracy as *the premise that the mass of citizens will make right decisions most of the time in response to critical issues.* For this to be true the citizens must be well informed. A great leader must be able to explain to the public *why* certain actions are right. Audience-response television is viable only if television meets its responsibility to inform the public well on critical issues.

Chapter Eight

SPIDERWEBS

Consider the technical progress that humankind has made in the last hundred years and reflect upon the accomplishments of the next hundred years, with hundreds of millions of people adding to the noosphere *linking millions of computers all growing explosively in power and intelligence.*

All round us, tangibly and materially, the thinking envelope of the Earth—the Noosphere—is adding to it internal fibers and tightening its network; and at the same time its internal temperature is rising, and with this its psychic potential.

PIERRE TEILHARD de CHARDIN

Pierre Teilhard de Chardin, the religious philosopher who reflected on the evolution of men, used the term *noosphere* (from the Greek root *noos*: mind). He visualized the earth enmeshed in a sphere of channels for the immediate interchange of thought, information, and intellectual operations. Today we are beginning to build networks with multiple paths like spiderwebs which interconnect computers and computer users. These networks are spreading worldwide and will become accessible from the home, possibly via a color television set with a keyboard added. While the networks are spreading, computers and information storage are dropping in cost at a phenomenal rate, and will continue to do so until powerful computers become as common as telephone sets. Large expensive computers will acquire extraordinary capabilities and will be accessible via the networks.

Whirlwind I, a pioneering computer of the 1950s, cost $5 million. Machines with the same computing power can now be purchased for $20. They are small mass-produced microcomputers. By 1980 they will be manufactured at a rate of many millions per year. All manner of different machines will have microcomputers hidden under their covers. As microcomputers proliferate, larger computers will become more powerful and acquire even larger storage units. For example, the Bible can be encoded into about 25 million bits; some larger computer systems store *one hundred thousand times* this much information. (A bit is one on/off condition in electronic devices, as discussed in Chapter 4.) A few years ago a person of average chess-playing ability who wanted to play with a computer needed the world's fastest computers; now a chess-playing machine is sold by department stores.

Despite their amazing capability, today's computers are only an elemental beginning. The large machines will attain much more elaborate software and vast libraries of information. The era of "artificial intelligence" in machines has not begun yet. Mechanisms and "dialogues" will permit untrained users to communicate fluently with the machines.

There are endless reasons why the machines should be made capable of transmitting to one another. One machine might need programs or data that another machine stores. A computer in one factory might send an order to a computer in another factory and receive a confirmation and invoice. A computer in a branch office might need to update the files in the head office. Hundreds or thousands of times as much information can be sent in machine-encoded form as when human speakers communicate over the same link.

Perhaps more important, *we* will be able to dial the computers and communicate with them. In offices, shops, factories, and homes, there will be small machines designed to enable us to communicate with distant computers. We will be able to ask them questions, to interrogate enormous banks of stored information, to perform calculations, and to enter data that the computers will store, process, and act upon. More advanced applications include a new type of thinking in which the creative ability of the human user interacts with the enormous logic power of the machine and in which the user has access to the machine's vast store of data. This interaction can produce results that neither human nor machine could achieve alone.

Computer terminals will provide many of society's services and information sources. Airline reservations, travel agent services, hotel bookings, brokerage services, banking services, car rentals, and so on will be obtained by machines. The public will use machines to obtain money, theater tickets, airline tickets, train seat reservations, stock market information, to do off-track betting, to manage their bank accounts, and so on. A farmer or small landowner wanting advice about certain crops may use a terminal to call an information service run by a fertilizer manufacturer. The farmer will describe the crop, weather, and soil conditions, and a computer will indicate how to treat the soil. It will give types of products, quantities recommended, and prices. The terminal may be used to request delivery of the product.

Such a service will benefit many product manufacturers when appropriate data networks exist. There are many similar types of services.

The computer industry has gone through several separate phases in its development. Computers were originally built as mathematical calculators. In the 1960s they were sold in large quantities for automating clerical operations, working their way through batches of documents and records. Meanwhile, terminals came into use, along with storage units in which the data were always available to be read. Computers then became skillful assistants to people carrying out a wide variety of different tasks. They could provide information for bank employees, for factory management, for lawyers or patent agents; they could assist in text editing, in engineering design, and in formulating and solving complex problems. Some users could browse through vast quantities of information or instruct the machines to search for the data they need. Time-sharing services, in which many users could share one machine by telecommunications, developed. When computer networks developed, a user had access to not one machine, but large numbers of machines in different parts of the country connected by a data network.

In the 1970s microcomputers developed, so that time-shared links were no longer needed to do most calculations. These could be done on a pocket calculator or a small local machine. Telecommunications links to computers were needed to provide *information* or perform activities not possible on the small local machines. In many of its applications the computer has become a new communications medium, providing information and carrying out a dialogue with its users. As this capability improves, vast networks interlinking many computers will be used. A user will not necessarily know where the answers to his or her requests are coming from as computers flash messages around the network to obtain the required data or services.

CHAIN REACTION

In our society the production of services is much less highly developed than the production of manufactured goods.

Many professionals of great skill pool their talents to create the production lines that turn out cars or television sets. This combining of skills to mass-produce goods has resulted both in much more elaborate and better products than would otherwise have been possible and at much lower prices. In contrast, many services involve the skill of only one person—a physician, a news reporter, a repairman, or a teacher. An amazing quantity of different knowledge and skill goes into the production of a jet aircraft. Computer networks will eventually make possible *services* that incorporate a similarly vast quantity of different people's knowledge and skill. Services will be in short supply as long as they are in the artisan stage, but computer networks can mass-produce information resources of the highest quality, making them available to consultants, drug stores, ecologists, and individual householders. A network for teaching will one day have as much skill built into it as a jet aircraft now has.

Computers act as a storage battery for human intellect, and data networks provide the means of distributing the resource. As time goes by, more and more human intellect will be stored in the machines. It will become unnecessary, for example, to do many forms of mathematics by hand; a machine can integrate, differentiate, and solve the most complex equations. Mathematicians will go on to more creative pursuits, and as they do so the machines will become still more capable. Doctors, generals, architects, and scientists will likewise have the computer networks taking over their routine work, providing them with better information, and allowing them to concentrate on the more creative or human parts of their jobs.

In the research laboratories a variety of disciplines classed as "artificial intelligence" are steadily developing. Various types of intelligent behavior will be built into the computers and networks to assist users in finding information they want or in solving problems. The machines will steadily acquire more knowledge and more capability to manipulate it.

The word "symbiosis" is being used in a new way in connection with computers. Webster's International Dictionary defines "symbiosis" as "the living together in more or less intimate association or even close union of two dissimilar organisms." In the new use of the word, one of the "organisms" becomes a computer with its associated equipment. People communicate with the machine by means of whatever "terminal" device gives them the closest relationship. This might be a machine with a screen, possibly a domestic television, on which the computer flashes diagrams, text, equations, or numbers. People give the machine information with a keyboard or with a "light pen," with which they can, in effect, draw on the face of the screen or in other ways indicate their wishes to the computer. The potential of human intelligence combined with the best capabilities of computer networks will take decades to understand fully, let alone exploit, but of all technological advances it may be that which brings the most change to society.

We are moving into an age when intelligent people in all walks of life will need and constantly use their computer terminals; this will be a symbiotic age when the limited human brain is supplemented by the vast data banks and logic power of distant machines. Probably all the professions will have their own data banks and possibly their own languages. The nonprofessional will use home terminals for education, working out tax returns, computer dating, planning vacations, and just for sheer entertainment.

Before long this technology will reach a critical point in growth, when it achieves mass acceptance and masses of people will contribute. Vast numbers of computer hobbyists will have access to vast networks of program and data libraries. The systems will grow, multiply, and interlink. A worldwide network of computers will be available to us. We will find enthusiastic users in many fields adding programs and data to files for public use. Dedicated amateurs are beginning to write programs and collect royalties on them. The most popular programs will earn massive royalties, as best-selling books do. This incentive will cause the libraries of programs to grow vast, like libraries of books and literature.

Mass participation in programming will cause an ever-growing chain reaction. Computer-assisted instruction will employ thousands of talented teachers with professional technical assistance. And this is equally true in other fields. Technology is bringing the computer to the general public.

People everywhere will be able to participate in using and building up an enormous quantity of computerized information and logic.

To an increasing extent the computers themselves will be used to *create* programs, automating the development process. The reader might consider the technical progress that humankind has made in the last hundred years and reflect on the accomplishments of the next hundred years, with hundreds of millions of people adding to the *noosphere* and linking millions of computers, which are growing explosively in power and intelligence.

BURST TRANSMISSION

As mentioned in Chapter 4, dialogue between a terminal and a computer does not generally need continuous transmission as in a telephone conversation. Instead, *bursts* of data pass backward and forward with periods of silence between them. If the cable transmits at a reasonably high bit rate, the bursts will be brief. In a typical dialogue with a computer the bursts range from a few bits to several thousand bits each. The pauses between transmissions from an operator are often half a minute or so, whereas the burst is transmitted in a small fraction of a second.

The reader might think of a data network as being rather like a railroad network with engines running on it carrying bursts of data. The switches must be designed to switch the tracks fast for individual engines. Engines from many different sources and going to many different locations will be intermingled on the tracks.

Many of the physical transmission links which form part of the telephone system operate at 1.544 million bits per second, and the network might be built from these links. A typical burst of data would traverse such a link in less than one thousandth of a second, and the switching would take place at about that speed. This may be contrasted with telephone switching, in which the track between users remains switched between them for five minutes or so while they talk.

If the gap between the bursts of data from one person

operating a terminal is twenty seconds, it will be seen that the transmissions from thousands of persons can be interleaved. Such a network will provide a very low cost per user if thousands of users share it.

The bursts of data are sometimes referred to as "packets," and one technique of switching the tracks for each arriving packet is called "packet switching." Packet-switching networks have been built in several countries but as yet are not widespread and do not use fast digital lines (e.g., 1.544 million bits per second). Several extensions to packet-switching networks have been demonstrated; these include *packet radio* with which data can be sent and received by radio devices, including portable terminals, which could be built as small as pocket calculators. Packet-switching networks have also sent packets via communications satellites.

Sometimes it is necessary to send long messages over packet-switching networks. These may be sent as strings of packets which are reassembled after transmission.

Data networks that can send bursts of data fast and cheaply will become very widespread in some countries during the next ten years. It is desirable that the means of connecting to such networks be standardized and such standards are coming into existence. In the future it may become possible to connect a packet device to *any telephone extension*.

There are now several inexpensive devices on the market that could act as cheap terminals to send and receive packets. One massive application for such devices is for machines that deal with monetary transactions in stores and restaurants. These are discussed in the following chapter.

It is possible that the home television set may become the most common interface to the data networks as in the British Post Office's *Viewdata* plans. The networks will grow and interlink worldwide. They will use much higher capacity transmission links such as satellite circuits and the AT&T trunks which transmit 274 million bits per second. The information and programs will be provided by thousands of entrepreneurs everywhere, ranging from private individuals to giant publishing and newspaper conglomerates. Directory computers will assist the public to find what they want, and many highly specialized resources will be made available to users everywhere.

Chapter Nine

INVISIBLE MONEY

Money is merely information, and as such can reside in computer storages with payments consisting of data transfers between one machine and another.

One of the most intriguing new uses of telecommunications will change the way we make monetary payments. Today most major money transfers are in the form of paper—bills, checks, or the paperwork associated wtih a credit card transaction. The new method is called *electronic fund transfer,* EFT.

Thomas J. Watson, Jr., the President of IBM, foresaw the revolution in money in 1965 as follows:

> In our lifetime we may see electronic transactions virtually eliminate the need for cash. Giant computers in banks, with massive memories, will contain individual customer accounts. To draw from or add to his balance, the customer in a store, office, or filling station will do two things: insert an identification into the terminal located there; punch out the transaction figures on the terminal keyboard. Instantaneously, the amount he punches out will move out of his account and enter another.
>
> Consider this same process repeated thousands, hundreds of thousands, millions of times each day; billions upon billions of dollars changing hands without the use of one pen, one piece of paper, one check, or one green dollar bill. Finally, consider the extension of such a network of terminals and memories—an extension across city and state lines, spanning our whole country.

The money is in the form of *bits* passing unseen between one machine and another. Electronic money can be transmitted over a telephone line, relayed via a satellite, delivered to a computer on a magnetic tape, or transferred in any other way computers handle information. Needless to say, such systems must be designed with a high level of security so that the invisible money cannot be accidentally lost or changed if machines break down or errors occur in transmission. The money must be protected from ingenious new forms of electronic embezzlement. Bankers are convinced that the security mechanisms can indeed be made adequate.

Over the next ten years the nature of the payments mechanism in some countries will swing from being predominantly paper-oriented to being electronic at least in part, with vast quantities of financial transactions traveling over data networks. Some large banks are now planning private networks with tens of thousands of terminals. There are more than 1400 banks in the United States, and eventually they will be interlinked into nationwide networks for transferring money electronically. Many institutions other than banks handle money, hold deposits, and offer credit; the financial data networks affect all such institutions and present sudden new opportunities that will generate fierce competition in the money-handling business. Eventually in the United States many millions of transactions per day will be passing over the financial networks.

MONEY IS INFORMATION

Before societies used money, trade was carried out by means of barter. Later people devised systems of exchange in which certain commodities became standards of value against which all others were measured. In early societies these commodities had intrinsic usable value, such as wheat, cattle, or wives. Later, gold became a standard. Gold was rare, divisible, unattacked by rust and lichen, and beautiful enough to inspire poets. For millennia it has been the world's standard of value—fought over, traded, ornamented, stolen, and worshipped. Only in recent years have we gathered the effrontery to question the necessity of its role.

Paper money was invented in 1694, and at first it was regarded by many as a sinister banker's trick. Until this century paper money seemed respectable only if it was backed by an equivalent amount of gold in some banker's vault. Today the currency note is no longer convertible into gold. Currency bills used to say that a central bank "promised to pay the bearer on demand" in gold the value represented by the bill. Today we merely say "In God We Trust."

In relatively recent years, checks came into common use by the average person, replacing the need to pay with currency. In the late 1960s check and currency usage began to give way to credit cards, removing the payments still further from backing by gold or commodities. Today, what may be the ultimate payments mechanism—electronic bits flowing between computers—is gaining momentum.

EFT recognizes that money is merely a form of information. The dollar bills that pass from pocket to pocket have become merely a demonstration of one's ability to pay. If money is information, then that information can reside in computer storages and payments can consist of data transfers between one machine and another. EFT enthusiasts began to talk of a cashless, checkless society.

In reality, society will be neither cashless nor checkless for the foreseeable future. Rather, what has worried bankers is that the amount of paperwork associated with checks and credit cards is growing by leaps and bounds. One hundred million checks a day are written in the United States, and without automated fund transfer this number would double in the next ten years. While checks are expensive because of the paperwork costs, credit card transactions are even more so; their cost is approximately $.50 per transaction and rising. Electronic fund transfer offers a way to slow, and later reverse, the growth of paperwork.

The replacement of gold by paper money and of paper money by checks were each revolutionary in their day. Now we must become used to financial transfers occurring in the form of electronic pulses on a data link. The paperwork associated with the transaction will now merely inform us about the transaction, rather than represent the transaction itself. It will not have to be keyed into a computer. The eventual consequences of the simple idea of automatic credit transfer will be enormous. Vast random-access computer files in banks will hold full details of all accounts. As a transaction is entered into the system, transmitted data will cause the appropriate amount to be deducted from an account in one computer and added to an account in another. Eventually, the financial community will become one vast network of electronic files with data links carrying information between them.

FOUR TYPES OF EFT

There are four main types of electronic fund transfer representing successive steps towards an EFT society. The *first* involves transfers of money between banks, to carry out clearing operations. *Second,* there are transfers between the computers of other organizations and the bank computers. A corporation may pay its salaries, for example, by giving a tape or transmitting salary information to a bank clearing center, which distributes the money to the appropriate accounts. *Third,* the general public use terminals to obtain banking services. These terminals include cash-dispensing machines in the streets. There are a variety of such terminals with different functions. To operate the terminals, customers are equipped with machine-readable bank cards. In the *fourth* and ultimate phase of electronic fund transfer, consumers pay for goods and services in restaurants and stores by using their bank cards or similar cards provided by American Express, large retail chains, petroleum companies, and other organizations. Today's credit card devices (which create paperwork) will be replaced by inexpensive terminals that accept the new machine-readable cards. Thousands of such machines are already in use.

"EFT" thus refers to a wide variety of different computer systems, but in general the term has become synonymous with advanced new technical directions in banking.

Present techniques for making payments require much human labor. Credit cards have increased, not decreased, the quantity of paperwork and manual operations. Labor costs are rising, and it is becoming more difficult to obtain workers for dull, boring, but high-accuracy tasks. It has been estimated that the overall cost of using credit cards exceeds $.50 per transaction in the United States and that the cost of equivalent EFT transactions could be lowered to $.07.

Electronic fund transfer can make cash available to bankers faster, and time is money, especially with today's interest rates. About 30 billion checks per year are written in the United States, representing $20 trillion per year. An electronic fund transfer network could speed up the average clearing time for checks by at least one day and probably more. One day represents a float of

$$\frac{\$20 \text{ trillion}}{365} = \$54.8 \text{ billion}$$

savable by electronic check transfer. At 8 percent interest this means a saving of $4.38 billion. Often several days' float will be saved.

The number of credit card transactions is also high and instantaneous handling could again give a massive potential saving to the credit card companies. The total transmission capacity needed is very low compared to today's usage of telephone circuits and satellites. A typical transaction can be sent in less than 500 bits.

AUTOMATED CLEARING HOUSE

A bank clearing house takes checks drawn against many banks and allocates the funds appropriately. The *automated clearing house* movement in the banking industry is an attempt to create an electronic infrastructure that can reduce the labor in check clearing. This will both lower the cost of check clearing and speed it up. It will also enable banks to offer new services to their customers. For example, computers in some corporations deliver the payroll in electronic form to a clearing house, and from there the money is moved into the banks where the employees have accounts. There is then no need to print and read payroll checks.

In the United States the number of automated clearing houses is growing. Between these centers a telecommunications network will operate, transmitting many millions of transactions per day by the early 1980s. The National Automated Clearing House Association coordinates the development of the clearing house facilities, which must remain neutral to the competitive banking industry. The automated clearing houses and their network will have a vital role to play as EFT systems spread to the consumer level.

PREAUTHORIZATIONS

Many of the payments that are made by check are repetitive payments, the same sum being paid at regular intervals, or at least a sum that can be calculated well in advance. Such payments include rents, mortgages, local taxes, society dues, interest payments. Much work can be saved if these payments are made by *preauthorization* (the term *standing order* is used in British parlance). Once an instruction to make the payment repetitively is given to a bank computer, the payment is made without further paperwork.

The U.S. government handles some military payroll and many social security payments in this way. Some labor unions have discussed having workers paid *daily* by electronic means. The preauthorized payments may be handled by an automated clearing house or by a suitably prepared bank. Many bank customers would welcome a bank service that pays their rent, mortgages, society dues, and so on, without involving them in further paperwork.

The situation is slightly more complicated if the payments vary each time they are made. Dividends, like wages, can be paid automatically into customer accounts, and most customers, once they are used to it, welcome rather than resist this form of computer-to-computer payment. In a similar way, telephone companies and other utilities could send their bills directly to the bank clearing system. This, however, is a much more drastic step, because money is being taken *from* the accounts of individuals rather than added to them. Many consumers feel that they should have the option of not paying their telephone bills! Nevertheless, the majority would probably welcome an automatic bill-paying service.

If preauthorized payments of these types were fully used, the total volume in the United States would exceed 5 billion transactions per year. The only paperwork would be periodic statements informing the bank users what transactions had been made. The paperwork would be statements about the transactions, *not the transactions themselves.* There would be no writing or processing of checks, no rooms of keyboard operators laboriously and sometimes erroneously entering details of transactions.

GOING INTO THE RED

If money is *deducted* from customer accounts by electronic fund transfer, some of these accounts are likely to go into the red periodically. The consumers do not have quite the same control as when they can add up every payment they make with their checkbooks (although few today do so).

An essential aspect of EFT, therefore, is the ability for customers to have negative balances in their banks. The magnitude of the permissible negative balance would be set by a bank officer (or possibly by computers). The customer would be automatically charged interest on a negative balance. There are various forms of automatic negative balance in operation today.

Automated credit offers a constant temptation to overdraw. This might have great appeal to banks, which would make money on the interest charged. Customers, on the other hand, might resent losing the time delay on the transactions or losing their ease of refusing to pay. A wide variety of incentives have been devised for making EFT appealing to customers; these include discounts, cash-dispensing terminals in corporate offices, EFT terminals in corporate cafeterias, lower bank charges, and general ease of obtaining money.

WORLDWIDE FUNDS TRANSFER IS HERE

Systems for transferring funds between banks electronically are coming into operation not only on a national scale but also internationally. The first major international network is the SWIFT system.

SWIFT, the Society for Worldwide Interbank Financial Transactions, is a non-profit-making organization set up and wholly owned by banks in Europe, Canada, and the United States. SWIFT operates a worldwide network the purpose of which is to send money, messages, and bank statements at high speed between banks. The participating banks financed the system, and a tariff structure charges for its use on a per-message basis plus a fixed connection charge and an annual charge based on traffic volumes. The banks range from very small ones to banks with 2000 branches. Networks like SWIFT will soon be moving millions of transactions per day between banks.

BANK CARDS

In the mid-1970s a new wave of banking automation swept across America, triggered by the advent of machine-readable bank cards. These plastic cards have the size and appearance of a credit card, but unlike conventional credit cards they carry invisible data that can be read by a terminal. Previously such cards carried data encoded on two magnetic stripes. Now a third magnetic stripe is used on which data can be *written* by the terminal as well as read. There are some exceptions to the magnetic stripe technology, notably New York's Citibank cards, which contain a stripe read with ultraviolet light.

CUSTOMER TERMINALS

Using a bank card at an appropriate terminal, bank customers can inquire about the status of their accounts. They can deposit or withdraw cash, borrow money if it is not in their accounts, or transfer money between different types of accounts. *In fact, they can do virtually everything that they would previously have done by going to a branch of the bank, standing in a queue, and talking to a teller.* The interesting question arises: If *all* a bank's customers used bank cards and terminals, would the bank need tellers? Its operations could be designed so it needed only officers, who deal with situations needing human interaction and decisions. A bank could close some of its branches in expensive city streets and yet give its customers more convenient service, because the automated teller terminals are becoming located in stores, shopping plazas, airports, factory cafeterias, and office buildings. Furthermore, the customers could obtain cash or other banking services when the bank was closed.

Some customers have an initial hostility to banking by machine, but once used to its convenience, few want to go back to queuing in marble-pillared branches.

The prospect of doing away with the bank teller is revolutionary enough, but another implication of automated teller terminals threatens to play havoc with the entire structure of banking. Banking in the United States has traditionally been regulated by state and federal laws saying where a bank may have its branches. In 1974 the Comptroller of the Currency, who regulates banking activity, made the ruling that a remote terminal which customers use does not constitute a bank "branch." Following this ruling, banks rapidly started to spread their tentacles into geographic areas from which they had earlier been excluded. The controversial ruling was then challenged in the courts and partially reversed, but nevertheless it seems certain that the structure of American banking will change fundamentally.

A landmark court case ruled that the bank terminals in the Hinky Dinky supermarkets in Nebraska were legal and did not constitute branches. In a Nebraska supermarket an individual can elect to pay for $25 of groceries by switching the $25 from a savings account to his account with the supermarket. The transaction is completed without the use of currency or checks.

Automated teller terminals can be operated by organizations other than banks. Several savings and loan associations operate them. Consumer finance companies, large chain stores, gasoline companies, credit unions, and other organizations collectively extend more consumer credit than the banks. Many of these operate credit-checking terminals, and some have applied for permission to hold customers' balances, in which case their terminals could have most of the functions of a banking terminal. The banks, in other words, could face electronic competition from a variety of other organizations.

WE CAN HOOK YOU INTO THE WORLD

The 1974 ruling that a banking terminal was not a branch opened up the possibility of large banks developing nationwide terminal networks. Later regulation may prohibit that, because the competition could become too severe for small local banks. However, the bank customers who have become used to electronic banking certainly want to have its facilities nationwide. A New Yorker wants to be able to use a New York bank card across the river in New Jersey and across the country in California. The customer not only wants to be able to use the card in stores and restaurants but also wants to be able to obtain cash from cash-dispensing terminals.

Nationwide networks for checking a customer's credit already exist. Nationwide networks for transferring consumer funds and accepting the bank cards will be built. Hooking together many local systems of banking terminals is a problem not unlike hooking together many local telephone or railway companies in an earlier era. There must be national standards for the terminals that are used. A nationwide network may develop to serve many different banks. A small-town bank, like an American small-town telephone company, will be able to say to its customers, "We can hook you into the world."

BANKING FROM HOME

The simplest customer-activated terminal is the telephone. Some banks offer services that enable customers to make payments by telephone. There have been experiments in which some banks have attempted to automate banking from the home by using a pushbutton telephone and computer which responds with spoken words.

One of the costliest operations in the use of computers is the preparation or keyboard entry of data. To reduce costs it is desirable to persuade customers to enter transactions themselves in an electronic form. This can be done at customer activated terminals in banks or in the street. It could also very conveniently be done from the home on a seven-day, almost twenty-four-hour basis. Paying one's bills at home with a pushbutton telephone could be very convenient and appealing to many customers. Some American banking authorities believe that banking from the home will become a commonplace.

Catalog or mail order shopping could be done from the home telephone, using EFT. This is done today without EFT by one large store in Canada, using a computer system which speaks responses. In the more distant future, television selling may enable viewers to purchase goods and pay for them using a keyboard.

POINT-OF-SALE TERMINALS

A great deal of human drudgery will be saved when the payments made by consumers in stores and restaurants are entered directly into the banking systems instead of being made by credit cards or checks. The rapidly spreading bank cards are the vehicles for such transactions, and consumers have demonstrated in a few early systems that they like the convenience of paying with bank cards.

The term applied to extension of financial services to stores and restaurants is *point-of-sale* systems. Eventually there are likely to be 50 billion electronic point-of-sale transactions made per year in the United States. Much of the initial use of point-of-sale terminals is not for fund transfer, but for checking the consumer's bank balance so that checks can be cashed.

The cost of networks needed for point-of-sale fund transfer and similar uses are, as elsewhere in telecommunications, highly sensitive to traffic volume. If there are few transactions the cost *per transaction* will be high; if there are many transactions the cost per transaction will be low. The initial systems may therefore be difficult to cost-justify, but once such systems exceed a certain volume they will become highly economical and will probably spread very rapidly, as did the use of credit cards.

One essential to the future of point-of-sale EFT is the availability of mass-produced, inexpensive terminals. Some terminals now in use are simple and low in cost. It is interesting to reflect that a bill-paying terminal for the home could be very inexpensive. Such a terminal could identify itself to the bank computer and have no need for a bank-card reader—only for a keyboard that sends signals like a pushbutton telephone. Manufactured in quantity, such terminals could cost as little as pocket calculators.

Because of the volume sensitivity, point-of-sale linkage to banks is coming either from the very large banks or from corporative bank card groups. In many cases, however, it is not the commercial banks that are providing point-of-sale terminals. The banks face potential competition in this area from credit card companies, savings and loan associations, finance companies, and particularly from the large stores themselves. The large stores see point-of-sale terminals as carrying out functions other than merely the cash transfer. They can provide better inventory control and sales analysis, tighter credit control, improved cash flow, shorter checkout time, and hence less checkout staff.

A factor which will make the area highly competitive in some countries is what organizations handle the sums of money, gigantic in total, that consumers deposit and that are extended as consumer credit. The battle for these funds, each running into hundreds of billions of dollars, will be fierce and is very much related to what credit cards or banks consumers use.

CRIME

Eventually a high proportion of society's payments will be made with machine-readable cards. If appropriate security procedures are built into the systems, a criminal would be able to gain nothing by stealing a bank card. They could be made much safer than today's credit cards (and in some systems have been). One of the subsidiary benefits of an EFT society could be a great reduction in street robberies because pedestrians would no longer carry much money or credit cards.

Major costs are loaded onto today's economy from robbery, theft, and fraud. They are reflected in the price of insurance, the reluctance of big-city police to investigate minor robberies, the fear of walking in city streets after dark, the inability to obtain some types of insurance. (Insurance is impossible to obtain on many welfare check payments due to the risk, for example.) Electronic payments systems with tight security controls could do much to lessen this burden of crime.

The EFT networks themselves offer new opportunities for ingenious computer crime, and tight security controls need to be built into the systems. The only way to make the transmissions safe from wiretapping is to use cryptography, just as intelligence agencies send coded messages that cannot be deciphered. The technology does exist for making EFT networks sufficiently secure, although on some systems it may not be used adequately.

Successful crime in a computerized society will probably require, like other activities, longer training and a higher IQ.

PRIVACY

A cause of concern with electronic fund transfer, as in other advanced uses of teleprocessing, is that individual privacy may be eroded. An individual's financial history might be laid bare to government authorities. Some doctors rebelled against using on-line services for billing patients, presumably because they revealed too much to the IRS.

The on-line terminal services, while accepted with delight by some, are regarded with distrust by others. There is still a fear, if not always consciously stated, that the remote machine will lose information, make mistakes, or print out one's financial details for other people to see. However, the computers *can* be made trustworthy, superbly accurate, and secure. The computer files *can* be made private, but it will take time for these facts to be understood by the public, and it may require new privacy legislation.

Chapter Ten

INSTANT MAIL

Once we are able to have messages delivered almost instantaneously, the way we utilize them changes completely.

The total cost of mail delivery is gigantic, especially in North America. Americans are not only the most communicative people by telephone; they also receive the most mail. More mail is sent in New York City than in the whole of Russia.

The U.S. Postal Service is the world's largest nonmilitary employer. It spends $13 billion per year and is heavily subsidized by taxpayers. Manual mail delivery does not pay for itself. Much mail could be sent by telecommunications, and with large enough volumes the cost would be a fraction of that for manual delivery.

Amazingly, 70 percent of all *first-class* mail in the United States is generated by computers. Much of this is destined for other computers. It is printed, split into separate sheets, fed into envelopes, sent to a mail room, stamped, carried manually to a post office, sorted, delivered to planes and sent to another post office, sorted again, delivered to a mail room, sorted, distributed, and then laboriously keyed for entry into another computer. All this ought to take place electronically.

Much noncomputer mail, including handwritten letters, can also be sent electronically. Once we use electronics we should ask, What is the best form for a message? Does it need to be written at all? Once we are able to have messages delivered almost instantaneously, the way we utilize them changes completely. Instant mail will be fundamentally different from mail delivery by weary postmen.

NUMBER OF BITS

A typical telegraph message can be represented with less than 1000 bits. In the early days of telegraphy it was expensive to transmit 1000 bits. Today, speech is transmitted on digital telephone circuits at 64,000 bits per second.[1] A typical telephone conversation lasts around four minutes and hence requires about 15 million bits. Thus a telephone conversation is equivalent to about 15,000 telegrams. Compared with the telephone, messages sent by telegraphy *ought* to be cheap with today's technology.

A copper wire pair like those that exist in thousands under the streets of a city can carry 2 million bits per second.[2]

It could carry 2000 telegraph messages per second. In 1975 slightly less than 1 million telegraph messages were sent per day throughout the world. A simple pair of copper telephone wires, therefore, could be made to carry all the world's telegraph messages in less than ten minutes per day. AT&T's digital coaxial cable system could transmit them in one second.[3]

Telephone traffic has to be trasmitted in *real time*; that is, when a person speaks, his speech must be transmitted almost immediately. To achieve a good grade of service (i.e., low probability of a caller encountering a *busy* signal), there must be idle channels ready for immediate use. No real-time systems achieve 100 percent utilization of their facilities. Statistical calculations determine how much idle capacity is needed to provide a given grade of service. Furthermore, the transmission capacity must be designed for the peak telephone traffic. The traffic during the peak hour of the day is several times higher than the average traffic; the peak day is substantially busier than that of the average day. On a typical network the channels are idle 80 percent of the total time. Such is the nature of telephone traffic.

On any network with idle capacity, mail or messages could be transmitted. On a telephone network, whether a public network or a private corporate network, they could be sent in the gaps between the telephone calls. As new facilities such as satellite systems are designed, they should be designed to handle a mix of telephone, video, mail, and message traffic.

If a mechanism exists for handling non-real-time traffic in the gaps between telephone messages, a remarkably large capacity for mail and messages is available on today's transmission links. It is worth looking closely at society's non-real-time uses of information, to see how the unused transmission capacity could be employed.

If sufficiently large volumes were sent, electronic mail would be much cheaper than manual mail. The cost of electronic mail will drop substantially, whereas the costs of typing, addressing, delivering, receiving, opening, distributing, and filing paper mail are rising. Mail delivery is not real-time: We are happy if it is delivered an hour, or a day, later.

We write letters, leave messages, send telegrams, order catalogs, transmit batches of computer data, and request books from libraries. This information transmission has two important characteristics. First, it can wait until channels are not occupied with telephone or other real-time traffic. Second, it can be interrupted in the middle of transmission provided that the interruption is done in such a way that no information is lost.

DIGITIZED MESSAGES

It is possible to convert any type of message into a digital form for transmission. Different messages require different numbers of bits. Table 10.1 gives some examples of approximate message lengths when messages are converted to bits and compressed, ready for transmission or storage. We could send photographs of various levels of quality, black and white documents like the output of a copying machine, voice messages like those we leave on telephone-answering machines, typed memos, telegraph messages, computer transactions, bank checks, and so on.

Table 10.1

	Message type	*Bits*
1.	A high-quality color photograph	2 million
2.	A newspaper-quality photograph	100,000
3.	A color television frame	1 million
4.	A picturephone frame	100,000
5.	A brief telephone voice message (voicegram)	1 million
6.	A brief telephone voice message with complex compression	100,000
7.	A voice message of codebook words	400
8.	A document page in facsimile form	200,000
9.	A document page in computer code	10,000
10.	A typical interoffice memo	3,000
11.	A typical flip chart	1,000
12.	A typical computer input transaction	500
13.	A typical electronic fund transfer	500
14.	A typical telegram	400
15.	A typical airline reservation	200
16.	A coded request for library document	200
17.	A fire or burglar alarm signal	40

The brief spoken telephone message can be relayed, stored, delivered to its recipient, and filed, like a telegraph message. Unlike a telegraph message, it needs no special equipment; it can be sent from and received by an ordinary telephone. It needs far more bits than a telegraph message if encoded in a simple manner. However, it is possible by encoding the message for transmission in a more complex fashion to reduce the number of bits required by 90 percent.

Devices that transmit documents are called *facsimile* machines. Many of them are in use today for transmitting documents over ordinary telephone lines. The good ones give a document quality comparable to that of a Xerox copier and take between 1½ and 6 minutes to transmit the document.

Documents exist which digitize the facsimile transmission and encode a page into about 200,000 bits. The page can then be manipulated and stored by computers. Such devices will be employed in *electronic mail* systems. A page of text and numbers encoded like a telegraph message needs about one tenth as many bits as a page sent in digitized facsimile form, but it cannot contain signatures, handwriting, corporate letterheads, or diagrams, which *can* be sent in facsimile form.

Messages can thus be sent in a large or small number of bits. Take your choice. As satellites and other digital facilities spread, the choice will increasingly relate to convenience of the users.

WHO GOES FIRST?

On networks handling mutliple types of traffic it is desirable to have a priority system. Some traffic has a higher priority or rank than others. A message for a five-star general might be sent ahead of a message for a corporal even if the latter arrived first. In the simplest form there could be two priorities: real-time and non-real time. However, there are different degrees of urgency in the non-real-time traffic, so several priority levels may be used to help ensure a fast delivery for messages that require it. The system organization may be designed to permit the following categories of end-to-end delivery time:

1. Almost immediate (as with telephone speech).
2. A few seconds (as with interactive use of computers).
3. Several minutes.
4. Several hours.
5. Delivery the following morning.

When more than one message is waiting for transmission at any point, the higher-priority messages will be sent first. The fact that much of the traffic is not in the highest-priority category will make it possible to achieve a high loading of the transmission facilities.

CATEGORIES OF MAIL

Table 10.2 breaks down the U.S. mail by type. The asterisks indicate which mail could be sent by electronic means, and hence potentially by digital channels. A single asterisk refers to mail that could be delivered electronically to the end user. Individual households may not send nor receive electronic mail. They may have neither the equipment nor the desire to change their mail-sending habits. At some time in the future electronic mail will reach into consumers' homes, but this case is not included in Table 10.2. When government and business send mail to households, this mail could be delivered to the local post offices already sorted for delivery. All local post offices could have a receive-only satellite antenna on the roof and a high-speed facsimile printer. Advertising letters and promotion material have not been included as potential electronic mail, because they may contain glossy or high-quality reproductions. Some advertising letters could be sent by facsimile machines. Newspapers and magazines also have not been included, although there has been much discussion of customized news sheets being electronically delivered to homes.

On the basis of Table 10.2, 22.7 percent of all mail is potentially deliverable to end users by telecommunications, and 22.8 percent is potentially deliverable to post offices. In 1980 this will be a total of about 50 billion pieces of transmittable

Table 10.2 The composition of the U.S. mail, with an indication of which mail is potentially deliverable by satellite or other digital channels.

Type of Mail	*Percentage*
Individual households to:	
Business	3.8
Individual households	16.0
Government	0.4
TOTAL	**20.2**
Government to:	
Business*	1.8
Individual households**	3.8
Government*	0.6
TOTAL	**6.2**
Business to business:	
To suppliers*	3.9
Intracompany*	1.4
To stockholders*	0.7
To customers: Order acknowledgement*	0.2
Bills*	6.7
Product distribution	1.3
Promotional materials	5.4
Other*	6.2
TOTAL	**25.8**
Business to households:	
Letters:	
Bills**	10.1
Transactions**	1.2
Advertising	12.6
Other**	4.5
TOTAL LETTERS	**28.4**
Postcards:	
Bills**	0.7
Advertising**	2.1
Other**	0.4
TOTAL POSTCARDS	**3.2**
Newspapers and magazines	13.6
Parcels	1.3
TOTAL BUSINESS TO HOUSEHOLDS	**46.5**
*Business to government**	1.2
TOTAL BUSINESS	**73.6**

*Potentially deliverable by satellite to the end user (22.7 percent)
**Potentially deliverable by satellite, sorted, to a post office (22.8 percent)

mail per year in the United States. Some would require 200,000 bits each to encode; most would require less because alphanumeric encoding rather than facsimile would be used. The daily total would be roughly 3 million bits, the capacity of 543 digital telephone circuits[1]—a small fraction of the capacity of one of today's satellites.

If only 1 percent of such mail were transmitted, it would pay for a large satellite system. The practicality of satellite mail is aided by the fact that three quarters of all U.S. mail originates in only seventy-five cities, and only 20.2 percent of all mail originates from individuals, the rest coming from business and government.

FASTER-THAN-MAIL TRAFFIC

Today, message delivery that is faster than mail costs substantially more than mail. Bulk use of digital circuits[1] offers the prospect of *instant mail* being lower in cost than any form of manual mail.

Even if instant mail becomes very low in cost, most corporate communication users will probably still use the telephone because of its convenience, friendliness, and the cheapness of the instrument. The total number of telephone calls in North America is far higher than the total number of written communications.

However, here is another provocative statistic. In the United States only 68 percent of long distance calls and around 70 percent of local calls are completed.[4] The rest encounter busy signals, no answer, or equipment failures. On the completed calls, the called party is reached only 35 percent of the time. In other words, less than 25 percent of calls attempted reach their desired party. It is estimated that this wastes 200,000 man-years of callers' time, which at $10,000 per year is equivalent to $2 billion.

The telephone-switching computers could be programmed so that callers could leave a telephone message, which would be stored and sent when person who was dialed is available. The system would ring the called party periodically until it could speak the message. The called party could use the telephone dial to ask for repetition of the message, give confirmation of its receipt, or dial a response. The system may be designed so that a user can dial his stored message queue from any telephone, key in a security code, and receive the messages that have been left for him. If 10 percent of all business callers who failed to reach the individual they telephoned also left such a message, that would be several billion messages per year. Even this high volume is less than the capacity of one satellite.

It would be possible to send telephone messages for which simple responses are required. The receiver would dial the response on the telephone after a local computer speaks a message, and the computer would receive the response and deliver it. The voice answerback unit would inform the called party what form of response was expected.

MANY TYPES OF MESSAGES

Verbal telephone messages designed for machine delivery in a form rather like telegrams are referred to as *voicegrams.* Voicegram service could be valuable and widely used.

Telecommunications facilities, appropriately organized, would give us the ability to send many different types of messages: telegrams, voicegrams, handwritten letters, automatic communication between typewriters or copying machines, data sent between computers, and so on. New technology satellites could make the cost of such messages very low. A single satellite could carry all such messages in its continent. They could be sent on the unused channel capacity of a telephone satellite. They could also be relayed over any other transmission facility, such as the telephone network, cable television, CB radio or other mobile radio facilities.

USES OF INSTANT MAIL

When mail or messages can be sent in seconds rather than in a day or two, the service is likely to be used in new ways.

Documents or diagrams can be exchanged by two persons while they are on the telephone to one another. On the other hand, they may not telephone at all; they may converse on a typewriter-like terminal if this costs less than a long-distance telephone call. Callers may prefer sending messages to talking on the telephone when there is difficulty in making telephone contact with busy people.

Some users of data networks have terminals which were installed primarily for sending messages to one another. Communities of users have grown up bound by the ease with which they can interchange messages. As data networks become more widespread, this capability to send a message cheaply to another person's terminal will become more popular. Most offices and many homes will have terminals for sending and receiving messages. Such a terminal could be derived by a cheap addition to an already existing electric typewriter.

In the British *Viewdata* system the television set can be a means of receiving instant mail. When a message is sent to a subscriber a light will glow on his television set, at home or in the office. He can then display the message. A guest late for dinner could send a message. Parents could sent a bedtime message to children every night. A firm could broadcast messages to all employees. *When mail is instantaneous it can become a dialogue.* Two Viewdata subscribers could talk back and forth via their television sets.

With computers the system can put *categories* of persons in communication. You might be able to search for bridge partners, baby sitters, or UFO enthusiasts (of the Third Kind).

Cheap, fast message networks make it possible to hold meetings at which the participants are scattered across the world in their offices or homes. The participants send messages in a formal fashion instead of making speeches. Not only are the participants in different towns or different countries; they may not all be attending at the same time. Each has a terminal that logs what the participants have said, giving a serial number to each statement. An attender can wander into his or her office, read what has been contributed during the last few hours, and make a contribution of his own if so inclined. A meeting between world authorities, all busy doing something else at the same time, could continue for days.

The following is an extract from the minutes of a conference sponsored by Bell Canada with participants in various parts of the United States, Canada, and Great Britain. Each entry has a serial number, which participants can use to refer to it. It has the name of the originator, the date and time of origination (Eastern Standard Time), and then the message. It can be seen that sometimes hours may pass between one contribution and the next. The speakers in this extract were, respectively, in London; Montreal; Menlo Park, California; and Washington, D.C.[5]

(88) WILLIAMS THU 31-JAN-74 8:20 AM. RE 86, I VOTE YES, BUT WOULD LIKE TO SUGGEST IN ADDITION THAT THE DEVELOPMENT OF METHODOLOGIES TO ANSWER THESE VARIOUS QUESTIONS MIGHT MERIT A TOPIC SECTION IN ITS OWN RIGHT. OFTEN OUR ANSWERS ARE ONLY AS GOOD AS OUR METHODS. RE 78, WE HAVE BRUCE CHRISTIE WITH US AS CSG, HE WAS THE NRS PSYCHOLOGIST UNTIL OCT 1973, AND IS STILL IN CLOSE TOUCH WITH THEIR WORK. I CAN CALL HIM IN AT ANY TIME.

(89) KOLLEN THRU 21-JAN-74 8:44 AM. IN REGARDS TO (86), I AGREE WITH THE STATEMENT IN GENERAL BUT WOULD LIKE TO SEE MORE ELABORATION ON POINT 3 WITHIN THIS STATEMENT. I FEEL THAT AT LEAST FOR ME, THIS POINT IS SOMEWHAT AMBIGUOUS. IN ADDITION, IF THERE ARE OTHER PARTICIPANTS WHICH HAVE TROUBLE WITH ANY OF THE STATED AREAS OF DISCUSSION, PLEASE EXPRESS YOUR RESERVATIONS AT THIS TIME.

(90) JOHANSON THU 31-JAN-74 9:09 AM. I AGREE WITH 88 THAT METHODS ARE CRUCIAL TO THE DISCUSSION, AND WOULD SUGGEST THAT THIS BE INCLUDED UNDER TOPIC 3. I SEE THIS AS RELATING THE TECHNICAL AND SOCIAL DIMENSIONS OF THIS CONFERENCE. THIS CAN, OF COURSE, BE MADE MORE SPECIFIC AS WE GO ALONG AND HAVE CONSIDERED THE FIRST 2 POINTS IN THE AGENDA.

(91) MORGAN THU 31-JAN-74 9:18 AM. I THINK WE MAY HAVE PEOPLE TALKING ABOUT TWO THINGS HERE,

WHEN I USED THE WORD METHODOLOGIES EARLIER IT WAS IN THE CONTEXT OF RESEARCH. I THINK THE SAME MEANING IS INTENDED BY WILLIAMS IN 88. I AM NOT SO SURE THAT THE SAME MEANING APPLIES IN (90). CLARIFICATION?

(92) JONANSON THU 31-JAN-74 9:29 AM. RE 91 I ALSO MEANT METHODOLOGY WITHIN THE CONTEXT OF RESEARCH.

(93) KOLLEN THU 31-JAN-74 9:29 AM. RE 91 THE FORMAL AGENDA OF THE TRAVEL/COMMUNICATION TRADEOFF CONFERENCE IS AS FOLLOWS: 1. NEEDS OF TRAVEL AND WHAT TYPES OF TELECOM CAPABILITIES WOULD SUPPLEMENT OR SUBSTITUTE FOR THESE NEEDS. 2. HOW WE COMMUNICATE (E.G. FORMAT, CONTEXT, ETC.) SEE STATEMENT (62) FOR ELABORATION OF THIS POINT. 3. HOW CAN COMMUNICATIONS AND NETWORK TECHNOLOGY ENTER INTO THE PROCESS OF COMMUNICATING GIVEN THE UNDERSTANDING OFFERED IN POINTS 1 AND 2 ABOVE.

Chapter Eleven

INFORMATION DELUGE

Our cultural heritage is increasingly being recorded on video media. We do not yet have the means to make this accessible.

Certain libraries throughout the world, often university libraries, have prided themselves on keeping a copy of every book published. A few still do, but they cannot continue to do so much longer. It has been estimated that by the year 2040 there will be 200 million different books. To store these would require 5000 miles of shelves. The cost of storage and cataloging could be awesome. A card catalog like those used today would require 750,000 drawers.

Even if such a library could be built and operated, it would be of limited value to its users. A user would have to go on a major expedition to locate the shelf position of a book, and then the book might have been borrowed by another user.

Clearly, the information deluge of our age needs electronic media to manage it. The volume of papers, journals, reports, and other information is far greater than the volume of books and presents even more difficult organization problems.

The quantity of recorded information that people create began to grow after the invention of the printing press. It grew slowly at first. The Gutenberg press did not encourage new authorship, since it took years to set an existing book into print. Science and scholarship consisted of the study and restudy of ancient texts. However, the Renaissance brought new ways of thinking about the world, and slowly people began to acquire new knowledge. The first scientific journal appeared in the 1660s, more than two centuries after Gutenberg's invention. By 1750 there were 10 scientific journals, and from then the number was multiplied by ten every fifty years, the approximate numbers being as follows:

10 scientific journals in 1750
100 scientific journals in 1800
1,000 scientific journals in 1850
10,000 scientific journals in 1900
100,000 scientific journals in 1950

By the 1950s the growth of recorded information was referred to as an *information explosion.* The term may not have been apt, because an explosion quickly ends its violent growth. The electronic media of the 1960s and beyond were to multiply the recording of information in a way unimagined in earlier eras.

The number of scientific *papers* is increasing at a greater rate than the number of journals. The sum total of human *knowledge* was estimated to be doubling every ten years by 1950 and doubling every five years by 1970. In most fields of research, even one as old as medicine, more papers have been published since World War II than in all prior human history. Computers in 1960 had one instruction manual; by 1965 they had ten; by the 1970s more than one hundred. The total engineers' drawings of a jet plane weigh more than the plane.

The new technology of information processing and transmission has turned up just in time for the needs of scientific progress. Looking at the history of technology, one observes a number of fundamental inventions that arrived in one field of research just when they were needed to permit progress elsewhere. The thermionic vacuum tube, for example, arrived just in time to allow development in telecommunications. The first moon landing depended on a variety of newly developed technologies, and if any one of them had not been developed until a few decades later, the landing would probably have been impossible. We have reached a state now in human learning when the quantity of information being generated in industry, in government, and in the academic world would be entirely unmanageable without computers. The growth of information has no end in view, only greater growth.

Automated means of filing and indexing research papers, engineering drawings, and other types of documents is becoming essential and must be coupled with means of searching for and retrieving the required information. Telecommunications links will enable people needing information to search with computer assistance through what is available. In searching for reports on a given topic, the user will carry on a two-way dialogue at a terminal with the distant computer. The computer might suggest more precise categorizations of what the user is seeking. The user might browse through many data indexes, titles, and abstracts before finding what is wanted. Library systems will be made "public" so that a large number of people can use them, and a large variety of people can contribute to building up the data in the system.

Banks of data on many subjects will grow in the years to come. The biochemist will be enabled to check what reports have been written on a particular topic; the lawyer will have the mass of required literature electronically available; and the patent agent will be enabled to carry out a search in "real time." The *New York Times* morgue has been computerized so that writers may explore what articles on a given subject have appeared in that and other news media. Eventually, all manner of information sources will be connected to data networks, making them available to researchers and other users throughout the world.

LIBRARIES

Future libraries will store many books, papers, and journal articles electronically. Users will browse by means of a screen and a telecommunications link. Many users could have access to the same book at the same time, and they could have pages of it printed if they wished. There will be no card catalog. Instead, the computer will maintain indexes, that will permit rapid searching for items, using the terminal screens. Such libraries will contain any matter that today appears in print, and much that does not.

Electronic libraries can be accessible from anywhere with appropriate communications links. They can be used from offices where fast printing machines exist or from homes that have a device for displaying pages on the home screen.

The card catalog of many libraries is very simple. You can look up an author or a book title. A user wanting to find books that relate to a given subject probably will not know the titles of all such books, nor their authors. Computer systems now exist which enable users to search for the information they want. To make this search possible, the subject matter in a paper or book must be described in an abstract or set of subject headings or key words. This coding and classification is being done on much new literature. With its help, the user can search at a screen for reports that have been written on a particular topic. Large corporations use vast computerized information systems for accessing documents and reports. Similarly huge systems of legal information and cases are in use.

Computerized information retrieval is not limited to searching books and professional papers. Writers and historians often want to search past copies of newspapers. On some newspapers today the morgue (so called because it is usually dismal and in a basement) is microfilmed, and computers can automate the search for reports relating to given subjects. As we mentioned earlier, such systems as the *New York Times* information retrieval system could provide access not only to published news but also to reporters' original copy, most of which is not printed. This automated morgue uses similar techniques to systems for searching libraries, legal documents, patents, scientific papers, and so on.

These systems, accessible from around the world via data networks, can place vast quantities of information at our fingertips. They will not *replace* books. One of the joys of life for many people will still be to curl up with a good book. The new media will increase the *diversity* of information sources.

VIDEO INFORMATION

If all worthwhile information were in printed form, the data networks now coming into existence could provide access to it. Computers would enable researchers, reporters, civil servants, lawmakers, and ordinary people at home to find whatever information they needed.

Unfortunately, even such powerful facilities will not be enough, because a further information deluge of immense proportions is growing. McLuhanites insist that video information is in many ways becoming more important than print. The world produces several hundred hours of new television per day. In the year 2000 it will produce several thousand hours per day. In addition there are movies and an increasing quantity of educational courses on video tape or film.

Much of this is being preserved for posterity somewhere in the world.

The total volume of video information is huge. The medium used to record a typical movie could record several thousand books. How much of the world's video output ought to be made accessible via society's library systems? Certainly, much of it can be forgotten, as can much that is printed. However, some of each day's output is worth preserving and should be indexed and made accessible, like books. The quality of video education is steadily improving, but many works that should be regarded (and treated) as classics are locked up, entirely inaccessible, in the archives of broadcasting or film companies. To mention only one such case, a colleague of mine made a brilliant film on Michael Faraday for the BBC. It was shown once; one copy now exists, and even its maker cannot gain access to it. It will probably never be shown again until some means of access to video libraries comes into existence.

Great drama, the best films and comedy, and moments on television that achieve greatness (sometimes unexpectedly) are part of our cultural heritage. We do not yet have the means to make them accessible. History is being recorded on video media. Events, politicians, ordinary people, and the characters of future history books are being filmed every day. You have a vague idea of what life was like in Victorian England, but *your* image of it is probably different from *mine,* because you have read different books and interpreted them differently. Likewise, you may have a vague idea of why the United States declared war on England in 1812 but you probably have very little information about the mood of the times that led to the declaration. Persons in the future who want to know about our era will have available video images of the people, television shows that reveal their attitudes, interviews with politicians, documentaries, and programs of news commentary and argument. History can never be the same again.

History can, however, be distorted, and film is an extraordinarily powerful and persuasive means of distortion. Clips of film taken out of context can be linked together to give a totally false impression. A film can show the atrocities of one side in a war without showing the atrocities of the other. Freedom from distortion will require freedom of access to original films or programs. We have a tradition of freedom of access to books; the burning of books is a cardinal offense associated with only the most disreputable of authoritarian regimes. Film destruction does not yet carry the same stigma as book burning, but it could be more potent in distorting the knowledge of the past. The burning of films will be easier to accomplish, because there are fewer copies. Using film, distortion of truth can be vivid and thorough.

Information retrieval computers for film libraries could operate in a similar way to those for book libraries. In the future, libraries could be automated so that films or video tapes could be automatically loaded and transmitted. Persons searching the library might then view segments of tapes that meet the search criteria and could check to see whether these segments meet their needs. Persons assembling documentaries or history or current affairs programs would sometimes find this service invaluable and might be able to record and edit directly from the library transmissions. It would be possible to build video networks in the 1980s, linking film libraries, production studios, universities, or research organizations. The cost of such networks would be very high today, but it will drop to a reasonable figure with appropriate satellite systems.

With appropriate switching systems the television cables into schools could be linked to video libraries. Some are so linked today, with manual library operation. Local town libraries may have screens linkable to a national video network. Eventually, the television cables into homes may be switchable to video libraries, although this would require a much greater total channel capacity than exists today.

In discussing books, films, television, and libraries we are concerned with society's *access* to its information sources and records. As the information deluge increases, society will increasingly need techniques to select, edit, screen, and intelligently search the vast quantities of information.

There are many types of access. Today's television is limited, in that the viewer has a very low level of selection. There is a wider choice with print. Access can be improved by the use of video libraries. It can be improved further by linking such libraries into a video network that can be accessed from many locations, and eventually from the home. As the trend from print to video information continues, society will increasingly need improved video channels.

Chapter Twelve

RADIO DEVICES

Small is beautiful if the pieces are in communication.

The lacing of the world together will be made more complete with portable radio devices. These add another aspect to the society change portended by new telecommunications.

For many years science fiction has been full of small, portable, radio-operated devices like Dick Tracy wristwatch intercoms and paging devices that find James Bond no matter what he is up to. With the transistor, portable radio receivers suddenly become a reality—often a menace. Now, with mass-produced microminiature circuitry, more elaborate portable devices are possible. A miniaturized telephone keyboard *could* be worn on the wrist. There are wristwatches on the market that remember dates and act as pocket calculators. Some pocket calculators are programmable like computers. Some pocket walkie-talkies using this new circuitry give an amazingly clear and strong signal. Devices small enough to be worn on the wrist can contain more than 10,000 transistors today. (Transistor radios often contain 6 transistors.) Semi-conductor experts estimate that in the late 1980s a device that small could contain one million transistors.

CITIZENS BAND RADIO

In the mid-1970s the sales of citizens' band (CB) radio took off in the United States at a rate that took everyone by surprise, not least of all the regulatory authorities. CB is a service intended for short-distance personal or business radio communication, radio signaling, and the control of remote objects and devices. It is used by the public in large quantities for car and home radio-telephones and walkie-talkies. Prior to 1974 most CB radios were used by truckers. Now most of them are used in private cars for chit-chat between vehicles, emergency communications, avoidance of police speed traps, and pursuit of the opposite sex.

CB users have a vocabulary of their own, with words like *smokey bear* (police), *picture taker* (radar patrol car), *beaver* (girl), *shakey city* (Los Angeles), *pumpkin* (flat tire), *lettuce* (money), and *Uncle Charlie* (the Federal Communications Commission which regulates its use).

The range of a typical unit is about three to twenty miles, depending on the geography. It is possible to place conventional telephone calls from a CB radio by means of a special telephone adaptor at the base station. Suitably equipped CB sets have a telephone dial or keyboard. Regulations restrict any one conversation to five minutes; however, because of the crowded channels, most people limit themselves to much shorter bursts of speech. CB radio cannot be used for advertising, soliciting sales, music, or entertainment.

Most countries of the world have no CB radio. In Australia a substantial proportion of the public had CB radio when it was still illegal. In North America its popularity will probably cause its use to expand until most cars have a CB. To achieve this, better control of radio interference and allocation of more frequencies are desirable. The widespread use of interconnections to the telephone network would be of value. All the CB channels in use together occupy a range of frequencies which is only a tiny fraction of that used by one broadcast television channel. The television broadcasting frequencies could be used for CB.

The explosive growth of CB is an illustration of what can occur when a new use of telecommunications suddenly becomes fashionable. In 1974 few U.S. families had heard of CB. By the end of 1976 one in every sixteen families had a family license to use it, and new license applications were flooding in at the rate of half a million each month. Television had a similar explosive growth in the early 1950s.

MOBILE DEVICES

CB is only one of several mechanisms for speaking by radio.

Dispatching systems are used mainly for communication between fleets of vehicles. The public is familiar with the radios used by taxi drivers. Many vehicles share in the radio channel. An operator controls the fleet and can hear the messages from all drivers. In many systems, a driver can hear the operator's messages to all other drivers, although more elaborate addressing schemes can prevent this. Dispatching systems are used for controlling delivery trucks, police vehicles, fire engines, ambulances, military vehicles, and so on. Portable transceivers can be used in a similar fashion and are employed by police and security guards.

Radio telephones provide a switched two-way channel between two conversing parties, like a conventional telephone. The radio telephone channel is connected to the facilities of the public telephone network. Calls can therefore be placed to any telephone subscriber from a telephone in a car, and if the car is in a suitable location it can be called from a conventional telephone. Because of the shortage of radio channels, few people can have radio telephones today. In some cities there are more persons on the waiting list than subscribers. The few who use radio telephones are plagued with busy signals because often no channel is free.

There is a major new technology for radio telephones[1] that enables any one radio frequency to be reused many times and gives excellent sound quality even when the user is in a car speeding through city streets. Portable telephones that could be carried in a pocket or briefcase have been demonstrated.[2] In the United States in the mid-1970s, the Federal Communications Commission allocated a large block of frequencies for mobile radio use. It is therefore possible now to provide a radio telephone system that would allow hundreds of thousands of people in a large city to have excellent service. Such facilities will be built in some cities during the next few years.

Radio paging systems provide *one-way* channels to wandering users. The user carries a small unit clipped to a belt or in a pocket. Many such units share the same radio channel. Each unit reacts only to signals addressed to it, ignoring those to other units. The unit makes a beep to attract the attention of its wearer. Some units receive a voice message following the beep; some can only receive beeps. A paging user receiving only a beep goes to a nearby telephone. He may dial a paging operator who has a message for him or who connects him to another caller. Some private-branch telephone exchanges handle paging so that the person receiving a beep dials his own number with a paging code and is automatically connected to the extension that paged him. Paging receivers can be small; one is designed to be worn on the wrist. Some paging systems are connected to the public telephone network so that a person may be paged from any telephone.

Paging systems are used for locating executives, sending duty calls to volunteer firemen, and sending messages to representatives, newspaper reporters, doctors, and traveling staff. One of the most common uses is for maintenance and service personnel. When an office copier or computer breaks down, the local repair worker is contacted quickly by paging. Some such systems operate over geographic areas of fifty or one hundred miles. When paging transmitters are coupled to a nationwide network, it is possible to page a person anywhere in that nation.

SHORTAGE OF RADIO FREQUENCIES

The big problem with mobile radio devices is that the radio bands are overcrowded. The new micro-miniature electronics will permit devices with much more complex control mechanisms that can automatically search for free channels. The same radio frequencies will be used by many devices in different areas at the same time. This technique and the new allocation of frequencies will permit a proliferation of mobile radio devices in the United States. In some other countries there is little or no activity in this field.

AT&T and others have commented that the techniques they have developed can make the quality, operation, and grade of service of mobile telephones similar to that of conventional telephones. Motorola has demonstrated the ability to make portable telephones not much larger than pocket calculators.

DATA RADIO

The public normally thinks of radio as transmitting speech, music, or television. It can also transmit data. Transmission of data is much more economical than transmission of other information, and a large amount of information can be sent in a short time. The entire text of the Bible can be transmitted over one television channel in four seconds—much quicker than when Charlton Heston does it.

Data radio devices can be designed so that they only receive data or so that they both receive and transmit. Receive-only sets are cheaper and generally cause less problem with frequency allocation, because there is one powerful transmitter in an area rather than many small ones. A device that transmits as well as receives, on the other hand, is capable of more versatile computer usage. It is now possible to build inexpensive computer terminals which can be held in the hand and which connect to computer networks via radio links. The potential uses of such devices are endless.

BROADCAST DATA

A sound radio channel can be modulated to carry data and could operate inexpensive printers in the home or business and perhaps carry such information as Reuters news, the stock market ticker tape, sports reports, and weather forecasts. A television channel can carry far more information than a sound radio channel. With equipment of reasonable cost it can be made to carry several million bits per second. A terminal might receive blocks of a few hundred or a few thousand bits at a time, depending on the application. The channel could therefore *either* transmit different blocks to many different users *or* transmit a large number of blocks to every user, offering them a choice.

Of particular interest is the fact that a television channel can broadcast data *at the same time* as broadcasting television. A system developed by Hazeltine Research superimposes the data on the television lines without significantly degrading the picture quality. If one bit is imposed on each line, that gives a data rate of 15,750 bits per second. Other systems transmit data in the gap between the television frames. A television signal transmits the even lines on the screen, pauses while the scanning of the screen flies back to the top left-hand corner, then transmits the odd lines, pauses again, and repeats the process. The pauses are referred to as the *vertical blanking intervals.* There are sixty intervals per second in North American television and fifty in European. In Britain a standard exists for the transmission of data in the vertical blanking intervals. In each interval 720 bits are transmitted.

TELETEXT

The transmission of data superimposed on a television signal is known as *teletext.*

In Britain the BBC developed a service called Ceefax.[3] Britain's Independent Television Authority has a similar service and began operating in 1976 to a limited audience. They broadcast pages of information in color, for display on the television screen. The viewer has a small keyboard that permits selection of particular pages, just as one could turn the pages of a magazine. The pages may consist of news summaries, weather forecasts, stock market information, details about television programs, or general-interest magazine pages.

Ceefax is designed to transmit up to 800 pages on one television channel without interfering with the television broadcast. This quantity of data pours through the viewer's living room continuously. When he selects a page, the Ceefax unit selects that page, stores it in a small memory, and displays it on the television screen. While the viewer is looking at a page, it is stored electronically in a small memory in the teletext unit.

The contents of a given page need not remain the same. It could be constantly changing, like the stock market ticker tape or the news flashed in Times Square. The user could select one page, programmed to change, and watch screen after screen of news, advertising, or stock market reports.

PSEUDOINTERACTIVE CHANNELS

One-way transmission, for example data broadcast over the air waves, can be made to appear like interactive transmission if a large quantity of data is sent.

The user reads one screen and keys in a response. The response identifies another screen, which is then displayed. The response may be very short, for example, "1" or "2" meaning *yes* or *no.* The receiving device adds this response to a constant associated with the current screen to obtain the address of the next screen. In this way a user could carry out a dialogue with the receiver for such purposes as medical assistance, computer-assisted instruction, or interactive advertising.

The Ceefax service employs only two lines in each vertical blanking interval—that is, 100 lines per second. If the entire television channel were used, either in North America, Great Britain, or elsewhere, more than 15,000 lines per second could be employed. With more efficient page encoding, 1000 pages per second could be transmitted.

If such a system were designed for an average response time of five seconds, each page being transmitted every ten seconds, the system could make a total of 10,000 pages available on one television channel. This would be enough to allow highly elaborate dialogues to take place at a conventional television set equipped with an adapter.

The adapter that the consumer needs to pick up a service like Ceefax is built into some expensive television sets. Several experimental teletext services of different types are in use in various countries, some employing cable television. Many of the early transmissions were oriented to businesses rather than home subscribers. Some companies plan to transmit racing results and sports news to bars and betting parlors.

An adapter which enables a conventional television set to pick up broadcast data need not be expensive. It has been estimated that the Ceefax adapter could be marketed for about $300.[4] If a full television channel were used instead of merely the vertical blanking intervals, the cost should not be much different. The adapter would be built into a microminiature circuitry, which is rapidly dropping in cost. With consumer acceptance the cost could drop as that of pocket calculators has.

The television studio equipment used by the BBC for the Ceefax service was not expensive. It consists of a minicomputer with a file that stores the pages. Studio staff use screen units attached to the computer to compose, edit and update the pages. The computer is connected to the equipment that inserts the rows of data into the vertical blanking intervals in the television signal ready for transmission.

POCKET DEVICES

Broadcast data could be picked up by a pocket device just as broadcast sound can be picked up by a transistor radio. It is possible to design a wristwatch that tells not only the time and date but also the Dow Jones (or other stock market) average. The latter would be received by radio with a small antenna in the strap of the watch. The watch could receive the stock market ticker tape information. If it had a small keyboard (like the pocket calculator watch), it could receive individual stock prices.

TWO-WAY DATA RADIO

Pocket machines that can transmit as well as receive data have a vast number of potential applications.

One very appealing technology for this purpose is *packet radio,* which transmits and receives very brief "packets" of data. Packet radio devices have long been used with conventional computer terminals on some systems instead of connecting them to the telephone lines.[5] Experimental terminals small enough to carry in a coat pocket have been designed using the same transmission techniques.[6]

Packet radio devices use radio frequencies similar to radio telephones. A radio telephone needs a channel all to itself while it is in use, because telephoners speak continuously. Packet radio devices, however, send only very brief bursts, a few thousandths of a second in duration. An active user might transmit such a burst every half minute or so. Consequently, hundreds of active devices can share the same channel. Since only a few of the existing devices will be active at any one time, thousands of devices could be allocated to one channel in the same area.

Many thousands of packet radio devices might be operating in a city, sharing a hundred radio telephone channels. (The United States Federal Communications Commission has recently allocated frequencies that could be employed for this.) The devices may be a little larger than a pocket calculator, with a pull-out antenna like that on a transistor radio. When the user has entered a field, as one enters a number into a calculator, the device transmits a very brief burst of radio. Because the burst is so brief, a battery small enough for a pocket machine can power it. All over a given city these devices would be transmitting their bursts. The bursts would be relayed to different computers, possibly thousands of miles away on a data network, and the computers would respond.

The reader might imagine the transmission of a burst as an instantaneous flash of intense light like that from a camera flashbulb. The light is a different color for each of the channels—a hundred different colors. Hundreds of flashes occur every second throughout a city. If by accident two flashes of the same color occur at exactly the same instant, those transmissions might interfere wtih one another. If that occurs and the signals are damaged the machines will detect this and will flash again at instants that do not coincide.

In other cities throughout the nation the same is occurring. The flashes in one city do not interfere with those in other cities, but the information in a flash can be relayed over a data network from one city to another in a fraction of a second. Thus, there can be many millions of packet machines throughout the nation intercommunicating or communicating with distant computers. The technology and frequency allocations could be such that most citizens could own such a device or every vehicle could be equipped with one.

USES OF RADIO DEVICES

How could packet radio devices be employed?

The applications are almost as numerous as the applications of computers themselves. Individuals could send messages to one another. Pocket calculators could gain access to computer programs stored elsewhere. They could gain access to all manner of information sources and collections of data. They could be used for audience-response television, for placing bets and obtaining race results, for monitoring certain stock prices.

In addition, they could be used to transmit and receive from conventional computer terminals in areas where the local telephone channels are poor or nonexistent—especially in developing nations. They could provide hand-held terminals for persons who walk around a factory, for patrolling security guards, for railroad marshalling yards, for farmers or foresters, for doctors, nurses or staff walking around a hospital. Contractors or users of earth-moving equipment could contact their base computer. Air travelers or commuters on tedious train rides could use their terminals for such functions as balancing their checkbooks, arranging meetings, holding network conferences, learning French . . .

Two thirds of a million Americans already carry personal radio receivers in the form of paging devices. There are many applications of radio paging, and its use is growing fast. It is used for calling foremen or expediters in factories, locating buyers or sales staff in large stores, calling service personnel, calling auxiliary firemen and telling them the location of a fire, requesting roving executives to come to the phone, and so on.

Some paging systems, such as those within a hospital, are strictly local. Others cover a large area. Eventually it will be possible with the aid of switching computers to locate a person almost anywhere in the country, provided that his transceiver is switched on and working. This situation need not necessarily constitute an invasion of privacy, because he would always be free to switch off his transceiver or leave it at home. His home area computer might store the identifications of persons who tried to reach him when his transceiver was off the air.

In industry or government, key persons may be made permanently accessible by means of a transceiver they are not allowed to switch off during working hours.

As with other forms of telecommunications, paging may spread from business usage to consumer usage.

Packet radio could be used in utility meters for transmitting their readings periodically, or for monitoring potential earthquake or volcanic areas. Weather monitors, seismic monitors, and so on could be parachuted into forest areas.

Burglar and fire alarms could use packet radio, immediately transmitting details of any violations. Individuals could be made to carry radio cartridges in high-security areas. If a person was detected by a light-beam system or other detector and was not transmitting the correct code, the alarms would go off.

The portable transceiver would give people a means of immediately contacting the police. This might prove very popular in a society in which the growth rate of crime rivals the growth rate of electronics. The police computers now spreading across the country would instantly pinpoint the emergency call and dispatch a patrol car. The fire and ambulance services could be contacted equally quickly. One can imagine a fleeing criminal being tracked by calls from persons near the scene of the crime as patrol cars close in.

Radio devices could be used on every vehicle to identify it automatically. These could be used for automatically paying tolls or parking fees, for opening garage doors, for security control, and for enabling police to detect traffic violations or stolen cars. They could be used in mechanisms for lessening city traffic jams, or for obtaining assistance in parking. Packet radio could be used on trucks or delivery vehicles for sending messages, making inquiries, controlling the routing, or fleet scheduling. They could be used on public transportation systems for controlling buses or other vehicles. People could use them for calling taxis. There are many highway safety applications, including fog and crash warnings, ambulance control, and breakdown assistance. Vehicles, buses, or accidents could be quickly located by triangulation.

A keyboard in the car or in a briefcase could enable salespeople to transmit orders as soon as they were taken. Customers with problems could signal maintenance workers, who might be anywhere in a city. Maintenance workers could signal for help if they needed it. On the way home people could signal their spouse to have dinner ready. Newspaper reporters could type and transmit their copy from their car.

Uses of packet radio might be linked to automatic payment mechanisms. Automatic toll collection might be used to help regulate the flow of traffic into congested city centers. Parking fees may be similarly collected. When a packet radio signals some condition, a small payment automatically moves by telecommunications between bank accounts.

The transceiver could also be used as a personal information source. One could have access to one's own records stored in computer memory. One's secretary or spouse could store entries on one's shopping list or diary.

If large numbers of portable transceivers come into use, new types of applications will become available. Those new functions would in turn increase the market demand. As with radio, television, and the telephone in their day, and recently CB radio in America, a snowballing effect would set in until much of the public owned such a device.

They could be used in children's toys for a wide variety of purposes. Children could have access to libraries or computer-assisted instruction. They could be invaluable for military field operations where other forms of communication are poor. They could be used in graphic applications on screens that require bursts of transmission at a speed too great for connection by telephone lines. In general, radio can extend the numerous applications of telecommunications to areas where cables cannot be used—in cars, on ships, trains, and planes, in the city streets and fields, at the bottom of the garden.

An appropriate numbering scheme, like telephone numbers, would be needed if subscribers carried their transceivers from one city to another. The same is true for radio telephones. In one proposed scheme each transceiver would be permanently associated with its owner, and a network of directory computers would be used to find the whereabouts of a person who was not then in his or her home area.

Not all signals would immediately reach their destination. The whereabouts of the person called may not be known to the network, or the person called may have switched off his or her transceiver, not wanting to be disturbed. In such cases the stored-computer voice would inform the caller that such was the case.

Like all new technology, radio transceivers have potential for ill as well as for good. Nevertheless, it seems likely that the social benefits will override or greatly outweigh the disadvantages.

The world of the ubiquitous portable transceiver may be years away. It very much depends on government regulation of the radio frequencies. But in time it may be as important to our society as the telephone is today.

Chapter Thirteen

SATELLITE AGE

One satellite could have enough transmission capacity to provide every man, woman, and child in North America with a computer terminal.

The communications satellite is a very simple concept, as was the steam engine. When such new concepts appear, it sometimes takes a while for their potential to be understood. The potential of the satellite is so great that it will change the entire fabric of society.

To make full use of this potential, we will have to launch communications satellites much larger than those of today. This could be done fairly soon, because of the availability of the *space shuttle.* A typical domestic satellite today is six or seven feet long and weighs four or five hundred pounds. Because the cargo bay of the space shuttle is sixty feet long and fifteen feet in diameter, the shuttle could help launch communications satellites weighing many tons. Unlike today's rockets, the space shuttle is *reusable;* it can fly back to earth and land on an airfield ready for its next mission. In contrast, today's launch vehicles must be abandoned after each launch. Imagine having to throw away a 747-jumbo-jet after each flight.

Although the potential of communications satellites is so great, satellite technology is entangled in legislation and politics that could prevent its potential being realized. As long as that potential is not understood by the public and by lawmakers, it is likely that the public will not gain the benefits. Unfortunately, most of the world's lawmakers do not have a glimmering of the potential. Talking to a U.S. senator before the 1976 election, I found that he did not even know that a satellite could be used for telecommunications.

STATIONARY IN SPACE

A communications satellite is, in essence, a mirror in the sky: Radio signals are sent to it, and it sends them back. It is placed in a special orbit 22,300 miles above the equator. In that position it revolves around the earth in 24 hours—exactly the time the earth takes to rotate—and hence appears from earth to be stationary in the sky. The satellite remains in that position for many years, unserviced and unattended, although occasional adjustments are made to its position by firing small gas jets on board. The satellite generates the electricity it needs from sunlight; at that height, it is always in sunlight except for brief eclipses in the spring and fall.

The equatorial orbit in which communications satellites sit, apparently still, is over 160,000 miles long, so there will be many satellites above the earth, using different frequencies (because of radio interference problems) and forming a vast web that links together our machines, wall screens, electronic mail devices, and computers. Any one satellite is far enough away from its signals to reach almost half the earth, but some worldwide transmissions will travel over two satellite hops. One satellite can handle a large quantity of information; *for example, four appropriately designed satellites could handle all the long-distance telephone and data traffic of the United States.*

DROPPING COST

The cost of satellite technology is dropping remarkably fast. There are two components to the cost: that of the satellite and its launch, and that of the earth stations which use the satellite. In 1965, when the first commercial satellite was launched, the investment cost of the satellite was $23,000 per voice channel per year. By 1971 it was $618.[1] By the early 1980s it will be in the region of $60, and when the space shuttle is fully utilized it will be much lower. These figures relate only to the satellite and its launch (referred to as the space segment). Earth stations (the ground segment) have been dropping in cost just as dramatically. The first earth station for AT&T's Telstar satellite cost in the region of 50 million. The earth station for Early Bird, the first commercial satellite, cost more than $10 million. Earth stations have dropped in cost until now a powerful transmit/receive station for the domestic satellites in North America can be purchased for about $100,000. Stations that receive but do not transmit can cost only a fraction of this amount. NASA's ATS-6 satellite, launched in 1974, was moved along the equatorial orbit to India, where it beamed down television to thousands of villages that had never seen television before. It was designed so that the signals could be received with locally made earth stations costing $70 each.

In general the larger the satellite, the smaller and cheaper the earth station. The space shuttle will permit the launching of satellites that make earth stations cheap enough to sell in mail order catalogs.

Satellite transmission is dropping in cost as fast as pocket calculators are. We need to make sure that the public can benefit.

CHANGE IN PERCEPTION

The perception of the value of communication satellites has changed since the first satellites were launched. At first, satellites were perceived largely as a means to reach isolated places. The cost of lacing Africa and South America with conventional telephone engineering would be unthinkable, but satellites offer an alternative technology. Earth stations began to appear in the remotest parts of the world. Countries with only the most primitive telecommunications put satellites on their postage stamps.

As satellites dropped from their initial exorbitant cost it was realized that they could compete with the world's suboceanic cables. Satellites then had a part to play in the industrial nations, linking the continental land masses. The owners of the suboceanic cables took political steps to protect their investment at the expense of satellites, but soon more transoceanic telephone calls were made by satellite than by cable. Television relayed across the ocean by satellite became common, because the cables of the 1960s did not have the capacity to send live television.

COMSAT (the Communications Satellite Corporation) launched four generations of satellites in six years. EARLY BIRD, the first commercial satellite to retransmit signals from a fixed position in space, was followed by INTELSAT II in 1967, INTELSAT III in 1968, and INTELSAT IV in 1971.

When the first INTELSAT birds brought competition to suboceanic telephone cables, the domestic telephone networks seemed immune from the threat. The cost per telephone channel of the early satellites was high, and the

United States Communication Satellite Act of 1962 said that only COMSAT could operate satellites and that they could be used only for international transmission.

As had often happened before, technology changed more rapidly than the law. The first North American domestic satellite was launched in 1972 by Canada. It was originally thought of as a means to communicate with Canadians in the frozen North and was called "ANIK," which means "brother" in Eskimo language. However, it was soon realized that the ANIK satellites would provide cheaper long distance telephone or television circuits than those of the established common carriers. The Canadians could take advantage of the harmful U.S. regulations and market ANIK channels in the United States. Antennas were set up in the United States to use the ANIK satellites, and for their first two years in orbit these satellites earned a return on capital investment that was unprecedented in the telecommunications industry.

A flurry of legislation in 1972 resulted in the U.S. Federal Communication Commission's *Open Skies Policy,* encouraging private industry to submit proposals for launching and operating communications satellites. The first U.S. common carrier to take advantage of the Open Skies Policy was Western Union, which launched two WESTAR satellites in 1974; these were the first U.S. domestic satellites.

A price war ensued for long-distance leased communication channels. A leased telephone circuit from coast to coast via WESTAR had a fraction of the cost of similar channels from the terrestrial common carriers. It seemed clear that the price could drop further with more advanced equipment.

It became clear that there were major *economies of scale* in satellites. A big satellite could give a lower cost per channel than a small one. To take advantage of the economies of scale, satellites should be employed where the traffic volume was heaviest. Nowhere was it heavier than in U.S. domestic telecommunications, and so it began to appear, contrary to the earlier view, that there was more profit in domestic satellites than in international satellites.

Nowhere was this perceived more clearly than in Bell Laboratories, the research organization of AT&T, the world's biggest telephone company, and the birthplace of the first commercial communication satellite, TELSTAR. A Bell Laboratories study gave the alarming conclusion that a few powerful satellites of advanced design could handle far more traffic than the entire AT&T long-distance network.[2] The cost of these satellites would have been a fraction of the cost of equivalent terrestrial facilities. However, government regulations prevented AT&T from developing the satellites which it, more than anyone else, could make good use of. The field was left open for competition. A number of corporations, at first small ones, announced that they would operate satellites, and AT&T proceeded to spend many billions of dollars per year on expanding its terrestrial facilities.

For WESTAR users the perception of satellites had now become that of communication pipelines linking five earth stations in one country. A further perceptual change was to follow.

While corporations and computer users thought of the satellite as providing two-way channels between the relatively few earth stations, broadcasters or would-be broadcasters saw it as a potentially ideal way to distribute one-way signals. Television or music sent up to the satellite could be received over a vast area. If a portion of the satellite capacity were used for sound channels for education or news, a very large number of channels could be broadcast. The transmitting earth stations would be large and expensive, but the receiving antennas could be small and numerous. The Musak Corporation envisioned three-foot receiving antennas on the roofs of their subscribers' buildings. Satellites offer the possibility of broadcasting television to vast areas of the world that have no television today. If more powerful satellites could be developed and launched, television could be broadcast directly to the hundreds of millions of homes in industrial countries. Even with satellites of today's power, television could be distributed in hundreds of regional stations for rebroadcasting over today's transmitters or cable television systems.

In the United States two new television networks are spreading nation-wide using satellite relays. One is the Public Broadcasting Network, and the other is a cable television network that distributes Home Box Office for an extra monthly charge. Both are noncommercial networks which promise a major programming alternative to commercial television. Japan has more ambitious satellite plans and has a satellite designed specifically for television broadcasting. This will permit the use of inexpensive receiving stations that can be installed in individual offices and homes. It is easy to imagine Japanese satellite television receivers, employing a larger satellite, spreading across the world in the 1980s as Japanese transistor radios did in the 1960s.

Broadcasting is usually thought of as having one transmitter and many receivers. However, when a satellite is used for two-way signals, a form of broadcasting is taking place in which there are many transmitters. Each earth station is, in effect, a broadcasting transmitter, because its signal reaches all other earth stations, whether they want it or not. Each earth station, like a radio set, tunes in only to the signal it wants to receive.

Because of this broadcasting nature of satellites it is limiting to think of a satellite as a "cable in the sky." It is much more than that. A signal sent up to the satellite comes down everywhere over a very wide area. To maximize the satellite's usefulness for telecommunications, any user in that area should be able to request a small portion of the vast satellite capacity at any time and have it allocated to him (if at that moment there is any capacity free). Just as with a telephone network, any user should be able to call any other user when he wants. In contrast to the telephone network, it should also be possible to request different types of channels—television, high-speed data channels, bursts for electronic mail, and so on. To enable many users to share satellite channels in this way, ingenious control mechanisms have been built which allocate the channel capacity to users when they request it.

For the first decade of communications satellite operation, most of the capacity of the satellites was used for telephone traffic and television. The technology has evolved, however, so that in a sense satellites are much more powerful for the transmission of *data*. It now appears desirable for technical reasons to carry satellite telephone traffic, and possibly television, in a *digital* form. When the telephone voice is digitized in a simple manner, it becomes 64,000 bits per second. When it is digitized in a more compressed fashion, 24,000 bits per second is a typical bit rate. A typical satellite contains many separate relays (called *transponders*), each of which transmits a channel of many millions of bits per second. Many telephone calls can occupy one such channel. The satellites planned for launching by Satellite Business Systems (SBS) will transmit many channels of 41 million bits per second each.[3] This transmission rate can carry one television signal. A substantial quantity of data can be sent using these bit rates. Such bit rates make telephone voice appear expensive by comparison with data transmission as a means of transmitting information.

The potential power of satellites for the computer industry can be illustrated by means of a simple calculation. When a person uses a computer terminal, he does not transmit continuously at the full speed of the channel. As we described in Chapter 8, the transmission consists of short bursts of data with substantial pauses between them. Most dialogues with computers result in not more than 1.0 bits per second passing back and forth *on the average,* although at certain instants a much higher rate is being sent.

The RCA GLOBECOM satellite, launched in 1976, has 24 channels (transponders). Each channel can transmit 50 million bits per second. This gives a total possible throughput of 1200 million bits per second through one satellite; however, it would not be possible to achieve 100 percent utilization of this bit rate.

A conservative assumption is that 20 percent efficiency could be achieved (much lower than today's equivalent on well organized terrestrial lines). Even this efficiency would give a usable capacity of *240 million bits per second.*

The combined population of the United States and Canada is about 240 million. Let us suppose, for the sake of this illustration, that *every* person makes substantial use of computer terminals. If the average working person uses them one hour per day and the average nonworking person one half hour per day, the total terminal usage would be about 160 million hours per day. Let us suppose that in the peak hour of the day the usage is three times the daily average. The total data rate in the peak hour is then

$$\frac{160 \text{ million} \times 3 \times 10}{24} = 200 \text{ million bits per second}$$

In other words, *one satellite could have enough transmission capacity to provide every man, woman, and child in the United States and Canada with a computer terminal.* In addition (as we have done the calculation for the peak traffic hour), twice as much non-real-time data (such as electronic mail) could be sent over the same satellite at times other than the peak hour.

This calculation assumes, as does any such discussion of the power of satellites, that it is possible to organize terrestrial facilities for the satellite channels appropriately. In the above case, how does one enable an extremely large number of users to share the same channel without interfering with one another excessively? Technology that would make this possible has been demonstrated, but in addition to the technology, appropriate structuring of the corporations or organizations that provide the facilities is needed.

In spite of the satellite's power for data transmission, it would not be a sound business operation to launch a satellite solely for use with computers. A relatively small proportion of all the traffic that might be sent by satellite is computer traffic. To maximize its potential profit a satellite should be capable of carrying many different types of signals: real-time and non-real-time, voice, data, electronic mail, and video. For all these signals, the satellite should be regarded as a broadcasting medium accessible from anywhere beneath it, not as a set of cables in the sky.

To summarize, the perception of what a communication satellite is has changed, and different perceptions have been:

1. A means to reach isolated places on earth
2. An alternative to suboceanic cables
3. Long-distance domestic telephone and television links
4. Television and music broadcasting facilities
5. A data facility capable of interlinking computer terminals everywhere

6. A multiple-access facility capable of carrying all types of signals on a demand basis

The changing perception of satellite potential has been related to the dramatic change in satellite cost.

DISHES ON THE ROOFTOPS

Some satellite users of the future will have small dish-shaped antennas on their roofs or near their buildings, pointing at the satellite. In some cases these will be small (say, four feet in diameter), for receiving signals; in other cases they will be larger (say, fifteen to twenty feet in diameter), for receiving and transmitting.

To deploy vast quantities of satellite antennas in this way, it is necessary to ensure that the transmission does not interfere with existing radio links on earth. Unfortunately, today's commercial satellites operate at the same frequencies as the telephone companies' microwave radio links. Therefore, a new generation of satellites is planned, operating at higher frequencies which are not used by any earth-bound facilities.

Many different corporations share the same satellite channel. Satellite transmissions can be picked up over a much wider area than the telephone companies' microwave radio transmissions. It is necessary to maintain a high level of security over some transmissions so that they can be received *only* by the party they are sent to. Extremely tight security can be achieved by enciphering the signals as wartime spies do with their transmissions; however, now the enciphering would be incomparably safer because it would be done by computer-like devices and the codes are virtually uncrackable. A computer can scramble the bits in such a formidable fashion that no cryptanalyst could unscramble them.

Today's satellites are an elementary beginning, like the first steam engine. One day we will look back at WESTAR, the first U.S. domestic satellite, which we use today, with the amused but admiring sense of history that we have when we look at Explorer I, America's first, grapefruit-sized response to Sputnik. The circular orbit 22,300 miles above the equator will contain masses of hardware. Beams of higher frequency and higher information-carrying capacity, possibly laser beams, will eventually be used by the satellites.

A major factor in satellite design is weight. If the satellite is heavy, it can have more power and more capability, and the equipment on earth which uses it can be cheaper.

The power of the satellite depends on how much electricity it generates from sunlight. Typical domestic satellites today generate 150 to 300 watts—enough to light two or three household lightbulbs. With larger solar panels (like those of the Skylab manned orbital laboratory used by the astronauts in 1975), many thousand watts of electricity will be generated.

There is no practical restriction other than cost to the amount of energy that could be generated in geosynchronous orbit. The total solar energy intercepted by a strip ten miles wide at the geosynchronous position is hundreds of times greater than the total amount of electricity consumed on earth. Indeed, there have been serious proposals from NASA[4] and Arthur D. Little[5] for building in the communications satellite orbit a solar power station generating about 10,000 megawatts—enough power to supply the whole of New York City. The biggest problem would be getting the power down to earth on a microwave beam sufficiently diffuse not to cause harm. The total cost of power generation in orbit could be lower if it is done on a large enough scale. Nuclear generators in orbit have also been proposed.

A second important factor in satellite design is the diameter of the dish-shaped antenna on the satellite, which points to earth. The power the satellite can relay is proportional (within limits) to the fourth power of the diameter of the antenna. For example, a ten-foot antenna will relay sixteen times as much power as a five-foot antenna. Five feet is a typical diameter for today's satellites. However, in the utter stillness of space, with no gravitational force or breath of wind, a large antenna can be a fine gossamer-like structure, impossible on earth. The 100-foot-diameter, aluminized mylar sphere of the Pageos satellite weighed only 120 pounds. NASA design studies have described antennas for astronomical telescopes ten *miles* in diameter, deployed in orbit as a fine spinning mesh pulled into position by centrifugal force.

FUTURE SATELLITES

The satellite uses described in this book are based on satellites similar to those which have already been launched. Satellites launched with the space shuttle in the 1980s can have more advanced uses, because they can have much greater weight, power, and antenna size. Unlike today's communications satellites, they might be assembled with the help of astronauts. Some future satellites will contain computers for switching and control purposes.

A design study done by the Aerospace Corporation for NASA describes wrist radio telephones that could place and receive telephone calls via the conventional telephone network.[6] They would not be much larger than today's digital wristwatches and would sell for about the same price. (The cheapest walkie-talkies today retail for $8.95 a pair.) They would use batteries that are recharged every night, and as with today's watches, all their electronics would be on one microcircuit "chip."

The wrist telephone would transmit directly to and receive from a satellite. In this proposed design the satellite weighs 7.3 tons and uses an antenna 200 feet in diameter. This can be of light-weight construction, because in orbit there are no winds, gravity, or vibration, no rain, frost, or corrosion—just the utter stillness of space. The satellite power requirement is 21 kilowatts, which is less than that used aboard the Skylab satellite in 1975. It is estimated that one such satellite could serve 2.5 million wrist telephones across the United States and that the cost would be $300 million.[7] The total value of the terrestrial U.S. telephone plant at the time will be over $100,000 million.

The same study describes even more ambitious satellite uses, made possible by more complex satellites, some with antennas two miles long. Satellites could be used for thorough surveillance of national borders, for aiding in more complete weather forecasting, pollution control, and energy use monitoring, for keeping track of radioactive materials, for tracking vehicles, railroad cars, or packages. They could provide wristwatch navigation aids that would tell the wearer's exact location and calculate the distance and direction of any given map reference; how accurately they did this would depend on the size and cost of the satellite. Satellites could provide relatively cheap data links between pocket calculators. Owners of such calculators could send messages to one another, use each other's programs, obtain stock market results, or retrieve information from large numbers of data banks.

Satellites are already providing secure worldwide communications for the military. They could give the police much better facilities than today and offer a means of tracking criminals across state or national borders. They could provide wristwatch emergency and rescue aids, burglar and fire alarms, or diplomatic hot lines between all heads of state.

Some time this century a vast industry will grow up, placing massive hardware in orbit in a ring around the earth 22,300 miles above the equator. Eventually, it will become economical to have service vehicles in this orbit, repairing, refueling, or assisting in the development of the satellite equipment.

That an immensely powerful communications technology is within our grasp seems beyond doubt. There remains the question of cost. Where will the money for the new thrusts that are now technically possible come from?

The total investment now being made on the development of *terrestrial* telecommunications facilities is huge. AT&T alone is spending $9 to $10 billion per year on capital improvement of the Bell System. AT&T's top management have indicated that they intend this level of expenditure to continue. The annual revenue from telecommunications in the United States is over $40 billion and is growing by about $5 billion per year. Much of the capital expenditure in the telecommunications industry is going into the trunks and trunk switching that satellites and demand-assignment equipment could replace.

Six ATS-6 satellites with appropriate transponders

(1974 technology) could carry as much traffic as the peak traffic on the Bell System toll network (i.e., not including *local* telephone calls).[7] If earth stations were associated with the toll offices in the 500 most populous cities, such a satellite network would cost less than $2 billion. If it had a lifetime of ten years, the cost would be $200 million per year. (Most telecommunications equipment has a forty-year lifetime.)

The next major thrust in the space segment should capitalize on the economies of scale that today's technology offers. This can be done only by a large organization with abundant funds. We will not see the low cost per channel that is now possible if we continue to launch small satellites like ANIK, WESTAR, and SYMPHONIE. Government regulation, however, is constraining AT&T from moving freely into the satellite business. NASA, which always launched the most advanced satellites, is being prevented from designing successors to ATS-6. The European administrations, while singing the praises of SYMPHONIE, are unlikely to put up American-launched satellites that would constitute major competition to European telephone trunks.

The barrier to acquiring low-cost satellite channels is thus not technological but political.

Chapter Fourteen

HOME

The telecommunications society will require far more creative people than today.

In the early computerized television games, bouncing balls or matchstick figures and spaceships moved across the screen and responded to the player's keyboard or joystick actions. With better electronics the images can become film sequences, and when the player shoots down a spaceship it explodes in vivid color across the seven-foot screen with flaming chunks of metal spinning past the player. Samurai warriors race across the screen and, if the player wins, die spectacularly. Kamikaze planes hurtle at the player or shoot-outs take place on jetport runways with quadrophonic sound.

These spectacular playthings use two synchronized film sequences compressed so that they are transmitted together over a single television channel. The player sees only one at a time. His game-playing responses cause a microcomputer to display text or still images on the screen and to cut at appropriate moments to one or other film track.

A wide diversity of such games will be produced and recorded on videotape. The videotape machine may be too expensive for most families to own and most might prefer to play with games transmitted on the television cables, where there can be endless variety. The machine which receives games can also receive utterly captivating education programs which the viewer interacts with.

On the Saturday afternoon of the future while the children play Star Wars on the screen, the head of the household may be immersed in sport. He watches one game on the big screen while a printer at his side spatters out news of other games. With his keyboard he can request the results of other games to be displayed on the screen. He can freeze a frame on the televized play at any moment and examine it. While displaying one game, the more affluent viewer may be recording a different one. With the recorder he can replay exciting moments.

The teenager of the family may receive messages via the television set from a video dating service. These tell him to watch the dating program on the local cable at a certain time when two of the potential dates the computer selected for him will be shown. If he keys in the correct password he can watch them being interviewed in the studio before committing himself. "Blind" computer dating always had its problems, but when the participants see each other on the screen before deciding to meet it is entertaining and often works out well.

The capability to mass-produce complex electronic circuitry is leading to many new consumer devices. Electronic calculators, television games, and wristwatches are a start. Microcomputers the size of a wristwatch can be mass-produced for less than $10. These, in combination with new communications networks, make possible all manner of new items for the home.

In the advertiser's paradise of America, all kinds of highly colored catalogs can become available on new telecommunications media, and there will be varied enticements for exploring them. Very elaborate presentations of products will become possible. The Sears Roebuck catalog, for example, might include film sequences, although the user will still be free to "turn the pages," to use the index, to select and reject. As with American television today, advertising would help to pay for the new medium. Perhaps critical consumer guides will also become automated to aid product exploration. Having scanned the relevant catalogs and inspected pictures of the goods in detail, the shopper could then use the same terminals to order items, perhaps with the money being automatically deducted from his bank account.

The telephone line, in addition to its normal use, could be used for activating household appliances. A family driving home after a few days' vacation or a person about to return from work may telephone home and then key some digits on the pushbutton telephone that switch on the heat or air-conditioning unit. This would save fuel costs. Before leaving for work a person will preprogram the kitchen equipment to cook a meal. He will then telephone at the appropriate time and have the meal prepared. Cookers with built-in micro-processors will become very elaborate, capable of accepting lengthy sequences of instructions such as "heat the oven to 450°F to cook the roast that has been placed in it; one hour later move some vegetables out of the freezer compartment; leave them to thaw for one hour; lower the oven temperature to 300°F; heat the vegetables in their aluminum foils to 300°F; switch on the dish warmer; heat the soup. . . ."

Fire alarms could be directly connected to the fire service and burglar alarms directly connected to the police. The reading of utility meters could be done remotely so that the meter readers would not have to visit the home.

Various think tanks and research organizations have produced long lists of services which modern telecommunications could provide in the home. Some experimental systems have set about providing them, proving that at least the technology can work. The following is a list of some of the possible home services, omitting ones that seem particularly dubious economically, such as menu selection and shopping list management. All the items on this list could be carried out using today's telephone system and multichannel television cables, preferably with reverse channels. Many additional applications would be possible if there were better video channels—an individual channel into each home, switched channels, or video channels *from* the home. As these would be substantially more costly, their existence is not assumed in this chapter, but again we may note the need for a society wired with channels of higher capacity then telephone.

Passive Entertainment
- Radio
- Many television channels
- Pay television
- Dial-up music/sound library

People-to-people communications
- Telephone
- Telephone answering service
- Voicegram service
- Message sending service
- Telemedical services
- Psychiatric consultation
- Local ombudsman
- Access to elected officials

Interactive television
- Interactive educational programs
- Interactive television games
- Quiz shows

Advertising and sales
Television ratings
Public opinion polls
Audience-response television
Public reaction to political speeches and issues
Television interviewers soliciting audience opinion
Debates on local issues
Telemedical applications
Bidding for merchandise on televised auctions
Betting on horse races
Gambling on other sports

Still-picture interaction
Computer-assisted instruction
Shopping
Catalog displays
Advertising and ordering
Consumer reports
Entertainment guide
City information
Obtaining travel advice and directions
Tour information
Boating/fishing information
Sports reports
Weather forecasts
Hobby information
Book/literature reviews
Book library service
Encyclopedia
Politics
Computer dating
Real estate sales
Games for children's entertainment
Gambling games (such as Bingo)

Monitoring
Fire alarms on line to fire service
Burglar alarms on line to police
Remote control of heating and air conditioning
Remote control of cooker
Water, gas, and electricity meter reading
Television audience counting

Telephone voice answerback
Stock market information

Weather reports
Sports information
Banking
Medical diagnosis
Electronic voting

Home printer

Electronic delivery of newspaper/magazines
Customized news service
Stock market ticker
Electronic mail
Message delivery
Text editing; report preparation
Secretarial assistance
Customized advertising
Consumer guidance
Information retrieval
Obtaining transportation schedules
Obtaining travel advice/maps

Computer terminals (including the Viewdata television set)

Income tax preparation
Recording tax information
Banking
Domestic accounting
Entertainment/sports reservations
Restaurant reservations
Travel planning and reservations
Computer-assisted instruction
Computation
Investment comparison and analysis
Investment monitoring
Work at home
Access to company files
Information retrieval
Library/literature/document searches
Searching for goods to buy
Shopping information; price lists and comparisons
Real estate searching
Job searching
Vocational counseling
Obtaining insurance
Obtaining licenses
Medicare claims

Medical diagnosis
Emergency medical information
Yellow pages
Communications directory assistance
Dictionary/glossary/thesaurus
Address records
Diary, appointments, reminders
Message sending
Dialogues with other homes
Christmas card/invitation lists
Housing, health, welfare, and social information
Games (e.g., chess)
Computer dating
Obtaining sports partners

WORKING AT HOME

Widespread use of home terminals may be sponsored by employers who expect to benefit from them. What would they use them for? The most obvious is work connected with the computer industry—writing programs, developing their own collection of programs for systems analysis, design, or other work, using complex design techniques programmed in distant computers, obtaining technical documents, looking up facts, running teaching programs, and exercising newly learned skills. They or their children may decide they need to learn a new computer language. For several evenings they must spend an hour with a manual and at a terminal using a teaching program. The only way *really* to learn a computer language is to program in it. So they will dial up a computer that enables them to program with a given language and then spend a period each evening developing their ability to use the new tool.

Given the right telecommunications *almost any white-collar work* could be done at home. Secretaries can type there or answer telephones. Typing pools can be distributed. Accountants can work with terminals at home. In writing a report, a group of authors can type the text directly into their respective home terminals. The report resides in the memory of a distant machine. They can then modify it, edit it, restructure it, snip bits out, correct each other's work, add to each other's ideas, and instruct the terminal to type clean copies when they were ready. Possibly magazine editing will be done with such aids in the future.

If the facilities are available it seems logical for some employees to spend at least part of their time working at home rather than at an office. The overhead cost of providing staff with offices and desks is very high, especially in large cities; some of this cost will therefore be saved by the use of home terminals. A manager can see what remote members are doing by telephoning them, dial up the computer they are using and examining their work, and sending them instructions on their home terminals. Indeed, it is possible that in the future *some* companies may have almost no offices. Some software companies for producing programs have cut costs significantly by having most of their staff work at home. Parents who must be at home to care for their children could also benefit from such a scheme; child rearing would no longer exclude them from having a challenging or well paid job.

COTTAGE INDUSTRY

The growth of computerized teaching—from today's experiments to tomorrow's industry—will need a tremendous amount of human thought and development of programs. Much of this endeavor may also take place in the home, just as books are written at home today. Step by step, a teacher can build up lessons on a home terminal. Occasionally he or she may dial a colleague to ask him to use *his* terminal to try out the lessons. When the work is near completion, the teacher may try it out in a classroom, study the reactions of students, and return home to make appropriate modifications.

This is possible not only for teaching programs but also for many other computer uses now within our grasp. Legal data banks, medical data banks, data banks for all types of professional users, and those for nonprofessional people seeking information are going to take an enormous amount of intelligent effort to build up. Much of this work is likely to be done with terminals, probably terminals that impose standardized formats on the data being entered. It is as though we have to rewrite all our textbooks and reference documents in a form that gives terminal users access to the data and instruct computers to search for and manipulate the information.

Here we see the beginning of a return to a form of "cottage industry." While some people are employed by corporations to work at home, others will do so on their own initiative, creating new teaching programs or software for us on the ubiquitous computer networks.

Enormous quantities of ingenuity and programming are essential to this new era—not the work of geniuses but ordinary step-by-step construction and testing. The work requires a high order of craftsmanship. In general, it is creative, enjoyable work, work that persons at home can do, work that disabled and in some cases blind people are now doing. It is work to which the hobbyist or the enthusiast who wants to make money in his spare time will contribute enormously. It is vital that the products of such work be covered by copyright laws, so that these individuals can earn appropriate royalties. Like the authors of best-selling books, the authors of programs or data files that come into widespread use will make big money from them. Some are already. If the right financial reward structure is set up the growth will be explosive.

HOBBY COMPUTING

One of the most rapidly growing pastimes is hobby computing. Computer hobby stores are springing up in cities around the world. For a few hundred dollars, sophisticated equipment can be bought and it is *rapidly* becoming much more powerful. Magazines for computer amateurs are rapidly growing in sales and symposia held for them attract massive crowds. Before long many schoolchildren will have their own computers, and young people will bring a new level of creativity to computing. From this mass of young enthusiasts many computer geniuses will emerge.

Computer hobbyists have more reason to *communicate* than other hobbyists. It is enormously fascinating to try out other people's programs. News about a particularly interesting program spreads rapidly and many people want to try it. Because of this it is natural that there should be a merging of amateur computing with ham radio, CB radio, Viewdata, and other forms of communications.

Industry is encouraging and enthusiastically selling to computer hobbyists. To really make the sales take off it should emphasize and lower the cost of every form of telecommunications access—radio, telephone, cable television, and data networks—so that the hobbyists have cheap access to all manner of programs and data, games, instruction, catalogues, and so on, and can communicate with one another. Computer hobbyists will have networks of unseen friends around the world adding to and trying out their programs.

Some hobbyists hope to produce and sell their own programs or make them available to other amateurs. Some are less creative and interested mainly in education: They will dial various instructional information-retrieval programs. Others are interested in playing games, doing puzzles, or indulging in mathematical recreations. The majority fall under the narcotic spell of programming. Working on ingenious programs (rather than the routine of commercial programming) is endlessly captivating.

Computer amateurs will have innumerable contributions to make to the development of this technology. Most technical fields are too complicated or specialized for amateurs to make names for themselves, but programming new computer applications has endless scope. To the old established computer professional the amateur computing magazines represent a new world, teaming with new and inventive things to do with computers. In every direction, new territories await the ingenuity and care of a dedicated amateur.

CONSUMER APPLICATIONS

Home users will in time have access to a wide variety of data banks and programs in different machines. They will be able to store financial details for their tax returns, learn French, scan the local lending library files, or play games with a computer. If they play chess with it, they will be able to adjust its level of skill to their own. Computerized news will be presented via interactive television; users will skip quickly through pages or indexes for what they want to read on their screens. Because the machine's files will be very large, foreign newspapers transmitted by satellite could also reside in local machines. Users may have a machine to *print* newspapers in the home, although use of the screen might be preferable. If they want back numbers, they will be able to call for them, using a computerized index to past information.

As illustrated previously, persons interested in the stock market could dial up a computer holding a file of all stock prices, trading volumes, and relevant ratios for the last twenty years. Possibly, stockbrokers would make the information available free and would provide analytical routines to their clients. When clients bought or sold stock, they would be able to give the appropriate orders directly to the stockbroker's computer via his own terminal.

On the other hand, someone lacking the money to buy stock could play at buying it. The computer would calculate the effect of his buy and sell orders, permit him to trade on margin, and pretend to give him loans, but no actual cash transfer would take place (apart from the cost of using the machine). When friends visit, he would be able to dial up his records to show them that he started with $100,000 and in six months had made $78,429. He could dream in glorious detail. At least he will have had lots of practice ready for when he does eventually become rich.

SHOPPING

Although a businessperson or executive might use a home computer terminal for scanning news items or for stock market studies, a homemaker is more apt to use it for shopping. In some countries, automated supermarkets have come into operation. The shopper at the supermarket picks up a card or presses a button for each item he wants to buy. He pays the bill, and the goods are delivered from the stockroom.

The advantages for the supermarket are that less space and less capital outlay are needed and there is less pilferage. Still, why must the consumer come into the store at all? One could scan a list of the available goods and their prices at several different shops on the home terminal and then use the terminal to order goods. The management of the organization can cut overhead to a minimum by eliminating stores and lessening their size.

The customer who buys through an automated catalog avoids the exasperation of fruitless searches for hard-to-find merchandise: out-of-print books, replacement bulbs for projectors condemned by planned obsolescence, spare parts for automobiles, rare phonograph records. There are tedious searches for special items like summer houses for rent, theater tickets for a particular show, or boats with a particular specification. The classified advertisements of the world will become available on the home color TV screen. They will be *far* more useful than the classified advertisements in newspapers because they will be so wide-ranging in scope and computer indexed. The user can state what he wants, however unusual, and the computers will try to find it. Do you want to rent a house on the Cornish coast for two months, find an au pair girl who cooks Japanese food, try parachute jumping, contact persons experimenting with computer controlled lawn mowers, buy a pet kookaburra?

Much of the work of agents of various types could be done more cheaply and efficiently by a data bank accessible from terminals. Until home terminals are common, it is unlikely that those who now make their living from putting buyers in touch with sellers will cooperate with each other to set up systems. Some newspapers and magazines, already competing with television for advertising revenue, may go out of business. Because printing unions are well organized to prevent the automation of printing itself, many newspaper publishers may be unable to avoid closing down when the classified advertisements emigrate to the television set.

The services just mentioned may be provided free by the advertisers; many others will have a charge, with the bookkeeping being done on a computer. An automated diary might cost $.01 per entry—a small fee for assurance that no appointments are missed and no birthdays forgotten. Hunting through dictionaries, almanacs, abstracts of literature, and similar sources might cost about $1 per hour, much less than a trip to a good library, and be more fruitful.

As the data bank and services grow, the market demand for terminals in the home will grow until every TV set will contain the necessary communications interface, possibly like the Viewdata system.

SPORTS

In top golf tournaments today, computers are used to keep track of everything that is happening. Observers stationed around the course report information to a computer station by means of walkie-talkies. The machine digests all the information, operates a scoreboard for the clubhouse gallery, press, and television, and displays on a screen hole-by-hole scores and such information as greens reached in par, number of putts on each hole, and lengths of drives on selected holes. Instantaneous comparisons between players can be produced, as well as all manner of asides, such as remarkable runs of birdies. And golf does not come up to baseball in providing comparisons.

The terminal owner of the future will presumably be able to dial machines giving up-to-the-minute detailed information on any sport instead of being restricted to the one or two items fed by the television channels. One imagines a fan of the future watching pro football television on a Sunday afternoon much as today, but with a small printer chattering away at one's side printing commentaries it has been instructed to give on other games. The sports viewer may request figures about other games on his television set whenever he wishes.

OFF-TRACK BETTING

Computers for handling off-track betting already exist in some cities. In some of these systems the public can place their own bets at terminals in public locations. It would be a natural extension of such systems for the public to place bets at home with a keyboard connected to their television set—the same keyboard used for other purposes. A viewer could observe the horses before a race, place a bet, and then watch the race. On a seven-foot color screen this can be as exciting as being at the track. If other uses of audience-response television were slow to take off, this would probably be highly popular with some viewers. It is more appealing than conventional off-track betting—and that took off like a race horse.

Many other forms of gambling could take place on the home terminals, with either real or imaginery money. This would include placing bets on simulated, rather than real, activities. During a plague of foot-and-mouth disease which stopped all horse racing in Britain a *simulated* version of the Massey-Ferguson Gold Cup race was run in a large computer. A mathematical model of the horse race was programmed with the help of racing experts, who provided the details of horses and their form over previous years, jockeys, distance between fences and so on. The computerized horse race met with the full approval of the National Hunt Committee and the Cheltenham Racecourse Executive. The BBC broadcast a full commentary on the race, with commentators sounding no less excited because the horses were not real. The commemorative Gold Cup was awarded to the winner.

Clearly, this concept can be extended. With terminals in the home, the racing enthusiast can have a race any time he wishes with real or simulated "off-track" betting. It could be a fine after-dinner entertainment. He and his guests could use the terminal to ask questions about the various horses' form. Great race horses of the past which never raced against each other, like Man o'War, Kelso, and Secretariat, might now compete in a computer-simulated fashion.

Many other such games will be played with the terminals. Who knows what forms gambling might take in the computerized society, with the home gambler's bank balance being automatically added to or depleted?

The time will come when the computer terminal is a natural adjunct to daily living. Soon computing will become a mass domestic market, and the computer manufacturers' revenues will soar. The airline industry, the automobile industry, telecommunications, and other complex technical industries all spent two decades or more of limited growth but then expanded rapidly when the *general public* accepted and used their product. This is now about to happen in the computer and data communications industries.

CHANNELS INTO THE HOME

Today there are four categories of electronic channels usable in the home in countries with advanced telecommunications:

1. *Telephone wiring.* A pair of copper wires goes into most homes, providing telephone service. A nationwide switching system enables any home to be connected to any other. The telephone cables can be used for data transmission, so any home could gain access to vast numbers of computer services.
2. *Cable television.* The television cables have thousands of times the capacity of the telephone cable, but they are not connected to a nationwide switching network. Hundreds or even thousands of homes share the same cable and so could not all be receiving or sending different signals at the same time. Television cables were originally designed for one-way transmission to the home; some newer cables also have a channel from the home, but this may be of low capacity.
3. *Broadcasting.* Radio and television are broadcast into the home. Data could also be broadcast into the home. One television channel could carry millions of bits per second of broadcast data. In some countries, there will be home reception of broadcasts directly from satellites.
4. *Two-way radio.* Citizens' band radio is used from some homes to communicate with persons on the move. Ham radio is used, with greater restrictions, to provide a worldwide network. Packet radio could provide inexpensive data communications to and from the home.

Missing from this list is a facility for switched video communications. Switched video networks would enable people to see each other, watch each other's children or video tapes, request information from video libraries, and obtain the medical services and educational facilities which we describe elsewhere (Chapters 3 and 19). The telecommunicating society will not mature fully until it has video channels that are switched like today's telephone channels.

Technologies for building switched video channels are now coming into existence. Probably in the future a new cable will be laid into homes—a glass fiber cable. Glass cables transmit light or laser signals down very fine, flexible, glass "hairs," which are referred to as *optical fibers.* The glass must be extremely transparent so that signals can be transmitted several miles through it without much absorption. To achieve this, glass of exceedingly high purity has been manufactured, and glass cables suitable for home wiring are now in experimental operation. The world's copper supply is running out even faster than its petroleum, so mass-produced glass fibers are likely to replace copper cabling. They are made from one of the world's most abundant raw materials—sand.

The signals sent through an optical fiber are one billion times higher in frequency than the highest frequency that can be sent through a telephone cable. In theory the optical fiber could carry one billion times as much information as a telephone cable. This greatly exceeds the capability of electronics so far, although much faster electronic devices are working in the research laboratories. The Japanese have marketed optical fibers that transmit more than one billion bits per second.

An optical fiber is very narrow, so one flexible cable the width of today's telephone cables, which are transported wound onto drums, could contain hundreds of thousands of such fibers. A future street in the suburbs will have a cable running beneath it, like today's television or telephone cable, with a separate pair of optical fibers going into each home.

The expense of wiring up all the homes in a country with optical fibers will be almost as great a proportion of the gross national product as wiring them with telephone cables was in an earlier era. In the United States it is likely to cost twice as much as the moon landing (in dollars adjusted for inflation) but the social benefit will be a million times greater. It will not happen overnight. Once it begins, it may spread quickly in some cities but perhaps take twenty years to reach remote rural areas. Indeed, it will probably never reach remote homes unless a law like the U.S. Communications Act of 1934 decrees that such a service should be made available to all people at a similar price. In 20 years time there may be a cheaper way of reaching *remote* homes—powerful satellites designed for two-way video operation, considerably more advanced and much larger than the satellites discussed in this book.

WIRED CITIES

Meanwhile there is much talk about "wired cities," or cities with switched video facilities into homes. To wire up a city with a high population density would cost much less per subscriber than wiring the suburbs or the entire nation. Still lower in cost per subscriber are wired building complexes. Multifunction wiring can link the various apartments, shops, and business premises in large city blocks, university campuses, shopping centers, or factory complexes. All can benefit if the facilities are planned in an integrated fashion.

A modern shopping center contains buildings that might include a supermarket, a bank, a restaurant, a post office, a broker's office, a travel agent, a real estate and insurance office, and possibly a local government office. When the owners of the shopping center plan its services they may also arrange for telecommunications services such as the following.

1. Telephone (with a common private branch exchange)
2. Credit authorization terminals
3. Bank card terminals
4. Cash-dispensing machines
5. On-line cash registers
6. On-line supermarket checkout stations
7. Piped music

Banking services for the entire center may be handled by the on-site bank.

8. On-line fire detectors
9. On-line burglar alarms
10. Police hot line
11. Closed-circuit television for security
12. Security patrol stations
13. Centralized guard console

All security services may be handled centrally.

14. Centralized reading of gas, water, and electricity meters
15. Computer control of heating, air conditioning, and lighting to minimize fuel bills
16. Computer terminals in the insurance office
17. Computer terminals in the broker's office
18. Stockbroker information services
19. Computer terminals in the post office
20. Airline reservation terminals
21. Other travel-booking terminals
22. Terminals in the real estate office
23. Terminals in the government office

All connected to distant computers.

A similarly long list could be drawn up for other buildings or sites such as office blocks, hospitals, hotels, or apartment blocks with shopping arcades, and industrial sites. There is much to be said for having common wiring, common switching or control mechanisms, common line disciplines, and in general for the sharing of telecommunications facilities.

The transmission facilities of the world have long been dominated by telephone requirements. There were telephone circuits within buildings, across cities, and spanning nations, and all formed part of an integrated telephone architecture. Today the situation is beginning to appear more fragmented, and separate technologies are emerging for the wired building, the wired city, and the wired nation.

POLITICAL PROBLEMS

In the United States regulatory problems could stand in the way of constructing the facilities that are best for a wired city. Today's regulation places constraints on both the telephone and cable television companies that would prevent either of them from constructing wired city facilities that would handle both telephone and television. Either industry would fight against integrated cabling owned by the other. There is a vast and growing investment in existing telephone and television cabling, and this investment cannot be written off quickly by commercial corporations.

Because of these problems it is possible that fully wired cities will not come into existence without government involvement. Government involvement can take a variety of different forms. A likely form is the implementation of pilot systems in selected areas, such as new towns or university areas. It may be difficult, however, to transfer the technology from a pilot system to nationwide reality. In spite of the technological leadership of the United States, it might be more difficult to achieve wired cities in the United States than in some other countires because of legal and regulatory entanglements.

PROBLEMS

As with many of the applications described in this book, the nontechnical problems may be more difficult to solve than the technical ones. The market to homes has major economies of scale. Only when a service can be sold to many homes is it profitable, and only when there are many services are householders eager to pay the cost of the electronics. Many of the services we have listed are likely to be attractive and profitable in a fully wired society, but not during the embryo phase when only a small proportion of homes are wired.

The marketing of interactive television may become easier, because many toys and teaching devices designed to be plugged into the television set are coming onto the market. The public will become accustomed to using a keyboard in conjunction with their television. Telecommunications may be a natural extension of the home market for microcomputers and cartridges.

Home markets for interactive television will grow because of hobbyists; they will grow because of education; they may grow because of fads devised by the cable television industry that become fashionable and sweep the country. The technology is available today, but setting up the services will take time, talent, and money.

Some consumer markets for telecommunications have taken off with explosive speed—for example, television in the early 1950s and CB radio in the mid-1970s. What the industry calls *subscriber response services* may take off suddenly, perhaps triggered by some fad. The ability to select any of the current top twenty pop records for home playing whenever desired might have great appeal. Off-track betting in the home could suddenly become fashionable. Watching the horses before a race on color television and being able to place a bet using the same set could have enormous attractions. Such a service might expand rapidly to betting on all types of sport, with the equipment installation cost subsidized in part by bookmakers and advertisers. Television game shows would probably join the bandwagon, and a bingo-happy nation would acquire the facilities that would later give it university courses in the home and encyclopedic information sources.

It will be a long time before the electronics industry produces robots for the home like those in *Star Wars*—even simple robots which can clear the table and keep the house clean. Instead something immensely more interesting and valuable can emerge—communications products which bring exciting education and information resources, and plug the home into the networks linking millions of computers which will fundamentally change work and travel patterns and vastly increase human capability.

Chapter Fifteen

TELEPHONES WITH PICTURES

Video links will be an integral part of the fabric of future societies, which will be confronted with rapidly diminishing supplies of gasoline.

Telephones have been built which permit you to see the person talking at the other end of the line. AT&T's Picturephone service was designed to give this capability, but it has not been very successful in the marketplace yet, because it is too expensive for most subscribers. However, future technology can greatly reduce the cost of Picturephone.

Sets that give a still rather than a moving picture, such as the RCA *videovoice* set, have also been marketed. This capability requires much less transmission bandwidth. Picturephone, like North American television, transmits thirty pictures per second to give the illusion of a moving image. To send one picture, say, every ten seconds requires one three-hundredth of the Picturephone's transmission capacity and could be done over a normal telephone line.

The Bell Picturephone has three components. First, there is a conventional *Touchtone telephone,* with twelve keys. Second, there is the *Picturephone* set with its screen, camera, and loudspeaker. Third, there is a *control unit* that contains a microphone and permits the user to adjust the picture and speaker volume. There is also a separate control unit, which is attached to a wall, often far away from the Picturephone set so that it does not clutter the user's room. The user does not need to hold a telephone handset while talking and can so talk naturally and is free to move.

Picturephone would be less expensive if a conventional mass-produced television set were used. Your friends could appear on the same screen as Kojak and Archie Bunker. Unfortunately, the cables in use are not yet good enough for this. The telephone network is too low in capacity, and the television network does not have adequate switching facilities or enough two-way channels. Consequently, today's Picturephone attempts to use wire-pair telephone lines, which is rather like trying to force a small river through a garden hose.

Three telephone lines are used, one for sound and one carrying the picture in each direction. The screen is small (5 inches by 5½ inches), and a black and white image is used. The set is an impressive piece of electronic engineering, but much of its expense arises in trying to overcome the limited capacity of the lines.

In a face-to-face conversation we look into each other's eyes part of the time. The Picturephone user will usually look at the eyes on the screen rather than at the camera lens and thus will appear to be looking away slightly. In order to minimize the annoyance of this, the camera lens is placed just above the screen, which is as close as possible to the eyes of the person on the screen. The displacement will make a user appear to be looking down slightly, as we often do in normal conversation.

The knob above the camera lens gives a distance setting. It can be set to 3 feet, 20 feet, or 1 foot. At the 20-foot setting it can view a room or a blackboard. At the 1-foot setting it views an object lying on the desk underneath it, thus enabling it to project documents or pictures. (In the early system installed at the Westinghouse Electric Corporation, one of the users used to put a photograph of the Westinghouse president in this position whenever the Picturephone rang). The viewing area under the set is 5½ inches by 5 inches. To view this, a swivel-mounted mirror automatically swings in front of the lens, and the picture scanning is appropriately reversed. The top of the field of view is the edge of the ring the set stands on.

The lens iris adjusts automatically to the lighting conditions. The strength of the video signal acts as an "exposure meter," but the upper and lower quarters of the picture are excluded from the measure to avoid false readings from ceiling lights or white shirts. The iris is a unique friction-free mechanism designed, like all of this equipment, to be as reliable as possible and not incur maintenance expenses. The set can be used in quite dim light.

The Picturephone call is initiated by the telephone set in exactly the same way as a telephone call, except that the # key is pressed prior to the keying of the telephone number. (The # key is the bottom right-hand one of the Touchtone keyboard.) The telephone number keyed will be the regular telephone number of the person in question. If a subscriber who does not have a Picturephone is called, his telephone will ring and an indication that he does not have a Picturephone will be displayed on the screen of the calling party. The calling party can talk to him and may comment on his lack of

Picturephone equipment. It was thought that this might put special pressure on subscribers in affluent areas to obtain Picturephone sets when they have become an accepted status symbol.

The telephone of the person called makes a distinctive Picturephone ring. When the person called picks up the telephone handset, the pictures appear on the screen. The user can adjust his picture with the SIZE, HEIGHT, and CONTRAST knobs. When doing so he can, if he likes, look at his own image on his screen by pressing the VU-SELF key. If he does not want to appear on the caller's screen, he can press the DISABLE key. He can mute his microphone, if he wishes, with the ON/OFF keys and can adjust the loudspeaker volume of his set with the VOLUME knob.

PICTUREPHONE PSYCHOLOGY

Picturephone has been heralded by AT&T as a major social innovation. Julius P. Molnar, executive vice president of the Bell Telephone Laboratories, wrote the following:

> Rarely does an individual or an organization have an opportunity to create something of broad utility that will enrich the daily lives of everybody. Alexander Graham Bell with his invention of the telephone in 1876, and the various people who subsequently developed it for general use, perceived such an opportunity and exploited it for the great benefit of society. Today there stands before us an opportunity of equal magnitude—Picturephone service.

He goes on to say in an article in the *Bell Laboratories Record*:

> Most people when first confronted with Picturephone seem to imagine that they will use it mainly to display objects of written matter, or they are very much concerned with how they will appear on the screen of the called party. These reactions are only natural, but they also indicate how difficult it is to predict the way people will respond to something new and different.

> Those of us who have had the good fortune to use Picturephone regularly in our daily communications find that although it is useful for displaying objects or written matter, its chief value is the face-to-face mode of communication it makes possible. Once the novelty wears off and one can use Picturephone without being self-conscious, he senses in his conversation an enhanced feeling of proximity and intimacy with the other party. The unconscious response that party makes to a remark by breaking into a smile or by dropping his jaw, or by not responding at all, adds a definite though indescribable "extra" to the communication process. Regular users of Picturephone over the network between the Bell Telephone Laboratories and AT&T's headquarters building have agreed that conversations over Picturephone convey much important information over and above that carried by the voice alone. Clearly, "the next best thing to being there" is going to be a Picturephone call.

Not everyone shares this enthusiasm. The London *Economist* in a special issue on telecommunications described the use of the Picturephone set as "a social embarrassment" and said that "talking into it was like talking to a mentally defective foreigner." A more common reaction is that users like it and enjoy playing with it, but are horrified when they hear the price.

Use of the Picturephone does create a feeling of closeness between the parties which is absent on the telephone. There is a slight distortion, especially when the face is too close to the camera, which is disturbing to some users. The camera is sensitive in the infrared part of the spectrum, and the human skin is partially transparent to infrared. This causes dark whiskers beneath the skin of a cleanly shaven man to be visible, especially when the light is dim.

When I talk to my wife on the Picturephone, the visual distortion seems to cause minor emotional distortion. The face talked to does not look quite as it should, and we tend to be self-conscious about the unattractive rendering of our own faces, which we can see by pressing the VU-SELF key. My wife thinks she needs Picturephone makeup and has a tendency to pull faces. Nevertheless, the call is more intimate and enjoyable than a telephone call.

Another disturbing aspect of Picturephone conversation is that callers tend to stare at each other's eyes. In most normal conversation people look only occasionally into the eyes of the person they are talking to. Lovers and salesmen may stare more constantly. Normally there is an unconscious but elaborate eye ritual during conversation. A new Picturephone correspondent tends to stare at you like a television announcer, and this sometimes makes conversation uncomfortable. After extensive use of Picturephone people become accustomed to it and more relaxed with it. Staff at the Bell Laboratories who have had a Picturephone for some years are remarkably casual in its use, often glancing at the set only occasionally. One man has a habit of initiating calls standing up, revealing only his torso to the party he calls.

A major tenet of Picturephone development has been to take maximum advantage of existing Bell System facilities. For this reason a conventional telephone is used to establish the calls. The control of the network and the signals used for dialing are basically the same as for the conventional telephone. What must be added are wires to carry the picture into the subscriber's premises, new switching facilities operated side by side with the telephone switching but under the same control, and trunks capable of carrying the Picturephone signals as well as today's telephone speech. All these factors have been worked out in a fashion that minimizes the consequent upheaval in an existing telephone plant.

EXPENSE

Any form of telephone service with moving pictures is likely to remain expensive until new types of transmission facilities are established. It requires a transmission capacity one hundred to one thousand times greater than telephones. If cities were wired with optical-fiber cables rather than copper wire cables and large satellites were in use, video telephone service using existing television sets need not be much more expensive than today's telephone service. But even in rapidly advancing countries that is two decades away.

For at least two decades video telephones will be expensive, and their use in homes and offices is likely to be restricted until new communications highways are built. Today, a few offices have Picturephones and a few have television transmission. In certain cases video links have proven very valuable.

In much of India corporations do not have telephones in everyone's office. Instead there is one room with telephones that many employees can use. Often executives have to make a booking in advance before they can use the telephone. The same will be true with video telephones in rich countries in their early years. Before it becomes economical for individuals to have video sets and dial-up facilities, there will be centralized shared sets and video-conference rooms.

The video connections will often be private, leased channels, sometimes with private switching. A few organizations have these today.

MEETINGS AT A DISTANCE

A video-conference room in industry or government may be designed as a facility in which persons can have meetings with other persons in a different location.

In one AT&T video-conference link, up to nine people sit at a curved table at each conference room location. Three microphones and cameras are used, with up to three persons within range of each. A fourth overview camera has the entire table of participants in view. A fifth camera, which is mounted on the ceiling and points at the center of the tabletop, is used to transmit documents, handwriting, small objects, or drawings, possibly with a speaker's finger pointing out features on them. Cameras can be placed in a variety of other positions, and often a camera giving a closeup of a speaker's face is used.

During normal operation, automatic switching will occur between the cameras. The logic of camera switching has been designed to imitate as closely as possible what a person does in a face-to-face conference. When a person speaks, the camera trained on his face is automatically switched on. The participants at the distant location will thus see his face on the screens in front of them. When another person speaks, voice-activated switching will switch the transmission to his face. If nobody speaks, the camera will remain on the last speaker for several seconds and the transmission will then be switched to the overview camera. It is rather like automated movie editing.

The screen in front of each participant normally displays the image being transmitted from the distant location. If someone is speaking, then the image will be that of the speaker's face. If no one is speaking, it will be the overview picture. When a person is speaking himself, he will thus see the entire group at the far location. When somebody responds there, he will see a close-up of that person. At the same time, the higher screen is displaying the image that is being sent to the far location.

The reaction of participants to AT&T's teleconference scheme was generally very good. One problem had to be dealt with: When persons coughed or sneezed, the loudness of this noise caused switching to the camera trained on them. One can imagine a conference during the annual flu epidemic of the locality, with the screens filled most of the time with the contorted faces of the various coughers and sneezers. The problem was solved by placing a "cough button" within reach of each participant. Pressing this button prevented the system switching to them when they coughed, sneezed or hiccupped.

For effective use in corporations it is often desirable for participants to exchange documents as well as to see and talk to each other. A telecommunications meeting room may therefore be equipped with facsimile machines for transmitting documents or drawings. Sometimes there is other equipment for special purposes. Video links are in use between certain hospitals and clinics, for example, and have been used to assist in detailed examination of a patient by telecommunications. Corporate and government facilities have included links to computer systems.

REMOTE USE OF DOCUMENTS

A major reason for needing video telecommunications is that documents can be read and discussed at a distance. It may be desirable to display them on the screen in such a way that both parties can point to them.

Unfortunately, today's television has inadequate resolution to display a page of typing, or the small print in a contract. Picturephone is even worse. Screen units that can display documents with high resolution are in use; they use two or three times as many lines to the screen as television. Future video telephones are likely to be available with high-resolution screens for *still* images.

High-resolution moving pictures would be too expensive to transmit, but such resolution is not needed for looking at a person's face. It would therefore make sense to design future video telephones so that the user can have one or more of the following options: a low-quality *moving* image like today's Picturephone, a high-resolution *still* image for document display, a means for the communicating parties to see each pointing at the documents, and a mechanism for printing the transmitted documents.

There are innumerable uses in society for telephones with pictures. Video links will be an integral part of the fabric of future societies, which will be confronted with rapidly diminishing supplies of gasoline.

Chapter Sixteen

A SUBSTITUTE FOR GASOLINE

The functions of a city will be different.

To make the West independent of the OPEC cartel, a vast sum would need to be spent on new energy over the next ten years. *Newsweek* estimated $500 billion.[1] The world's known supply of petroleum will run out in less than 35 years if consumed at the present rate. As it runs out it will be sought in increasingly inaccessible places, and its cost will rise rapidly.

While petroleum rises in cost, telecommunications will drop in cost and increase in capability. When substitutes for petroleum are discussed, telecommunications is generally not even mentioned. However, it does offer many kinds of substitutes for travel, and $1 billion would go a long way toward making them effective.[2] In the United States 53 percent of the petroleum used is for transportation. A study by the Office of Telecommunications Policy concluded that the use of automobile travel for acquisition, exchange, and dissemination of information uses 500 times as much energy in total as would be needed to do the same by telecommunications.[3]

Politicans declared in 1973 that they would make the United States independent of foreign oil within ten years. In reality, the United States is less independent now than in 1973. Energy *consumption* continues to rise, and oil imports have risen greatly. At the same time environmentalists have objected strenuously to the two main alternative sources of energy, coal and nuclear fission. It seems clear that the energy crisis will become much worse.

In a science fiction view of transportation substitution, individuals might stay permanently in their community, talking to each other on wall screens and operating factories and computers by remote control. Nothing quite so drastic is likely to occur, at least in the near future. There are, however, many different ways in which working and commuting patterns can be changed to save energy. Since they are likely to be used only if they also save costs, this chapter concentrates on facilities that would be cheaper than today's physical transportation.

Some substitutes for travel work efficiently with today's transmission links. Some require relatively minor additions to today's facilities. Others require a major expenditure on higher-capacity links permitting two-way video communication.

Some substitutes for travel operate with voice and data links or with links that transmit documents and still images. Some of them work with television broadcasting. It can sometimes be better to watch sports on television than to be there in person. If high-fidelity wall-screen television, quadraphonic sound, and photography with the best of today's lenses were used, then the excitement of the event could be brought to the living room and details of the play could be seen better than from a seat in the stand.

To simulate face-to-face human contact we need two-way video links. As we discussed in the previous chapter, the user may be able to select moving images of low resolution or still images of high resolution. Video links can be used for intimate, person-to-person communication as well as for conferences. A conference studio can have many screens, cameras on many participants, a special camera for viewing exhibits, and facilities for transmitting documents.

Lecturers and teachers may work in specially equipped studios to which people in many locations can be connected by videophone for live participation or by broadcast television links for passive participation. The students may be in their homes or in business locations. The lecturer can see the faces of the live participants, and they can see either his face or the diagrams, objects, slides, or film clippings he switched onto their screens. He can speak to any of the "class" individually, and they can speak to him. They cannot see their fellow students and so tend to ask questions with little embarrassment from possible class reaction. The teacher may occasionally let his students see some of the class. If he wants, he may switch the face of a questioner onto the class screens. On the other hand, he can address any one student without the others hearing.

COMMUTING

Perhaps the greatest waste of energy today is the commuting that takes people to work. It wastes not only energy but also a more precious resource—human time. Expensive equipment is wasted, because enough transportation facilities must be available to meet the rush-hour demand. The vast majority of commuters are white-collar workers whose jobs could easily be transferred from one location to another if suitable telecommunication links were available.

City planners of the future should be primarily concerned with moving information to people rather than moving people to locations where they can work with information.

There are several ways in which organizations could adjust to lessen commuting. First, the organization could move from a city center to the suburbs or to a rural area where employees can live nearby. Many organizations are indeed leaving cities like New York. In Britain the government has done much to decentralize government work locations, substantially lessening the commuting into London.

Second, the organization could split its main office, having many offices instead of one. The individual offices could be closer to where the employees live, and excellent communications facilities between the fragmented offices would eliminate the necessity for travel between them. An insurance company, for example, need not gather thousands of employees together in one large city building; instead, it could have many interlinked suburban or rural offices.

A third type of operation would be to have in suburban or rural towns facilities which several corporations share. The organizations might share the cost of teleconference and electronic mail facilities or of a small satellite earth station. In the future such facilities might be provided in office blocks which multiple corporations rent.

A still more decentralized form of operation, which would reduce commuting further, would be to have most employees working at home.

SECRETARIAL SERVICES

There are all manner of work activities that could be carried out from the home if suitable telecommunications devices were available. Many of them could be part-time activities for persons who look after children or for the semiretired.

Secretarial activities provide a good example. A secretarial pool could be composed of part-time secretaries working at home. A customer who needs typing done uses a dictating machine and transmits its contents over a telephone connection to an agency. Secretaries at home call the agency and ask for work. The agency catalogs the jobs and allocates them to employees. It transmits the recording by telephone to the secretary's home, where it is rerecorded. The secretary types the work on a machine that stores the typing on a cartridge or magnetic card. (It is easy to correct errors on such a machine.) The job may then be transmitted via the agency to the customer, whose machine prints or displays it. The customer may make corrections on a screen or computer terminal, or may indicate that corrections are needed and have the work passed back to the secretary. When it is word-perfect, the customer signs it. It may be sent by either electronic or conventional mail.

Here, both the customer and the secretary could work at home. They could also work in small offices near their homes which provide the word-processing equipment and which are connected to nationwide networks. Other secretarial services, such as keeping an executive's diary, could also be done remotely. There are many other types of jobs where the employees could work at home. Almost all white-collar work could be done at one end of a telecommunications link. The agency acts as a type of switching center interconnecting the users of a service and the persons providing it. Accountants, lawyers, programmers, and society's numerous paper-work shufflers could all operate in a similar fashion. Face-to-face contact is rarely needed, but where it is needed, the work may take place at a local office equipped with a video link.

Many corporations, especially the larger and more bureaucratic ones, seem opposed in principle to employees working at home. It is too much of a cultural change. They excuse their resistance by saying that employees are only diligent and productive where supervised and at their desk from 9 A.M. to 5 P.M. In reality, many employees succeed in doing remarkably little from 9 to 5. They manage to kill time until they can go home. They would do *much* more if they were paid only for results, and if paid for results they could equally well work at home or in a village office—with telecommunications. It would be cheaper than providing city office space for them.

It is time people started to ask in organizations everywhere whether it is really necessary to build larger and larger office blocks, with severe commuting problems, to house typewriters, adding machines, telephones, desks, computer terminals, and filing cabinets, when these facilities could equally well be in small local offices or at home. Electronics will do away with most paperwork; at the same time, it should do away with most commuting.

INTERCITY TRAVEL

After commuting, the second largest potential saving of transportation is in intercity business trips. Studies in the United Kingdom indicated "that about 40 percent of personal [intercity] travel could be replaced by audio conferencing [and] using video conferencing would add another 20 percentage points to the number of face-to-face meetings that could be replaced by telecommunications."[4]

The cost of commuting is paid by the commuter, but the cost of intercity travel is usually paid by corporations or other organizations. Thus, these organizations have a more direct financial incentive to lessen intercity travel than to lessen commuting. Many nationwide corporations in the United States spend as much money on airfares as on telephone service. Most corporations spend much more on car travel than on air travel.

Intercity telephone service is excellent in much of the developed world; therefore, if intercity travel is going to be replaced, it must be by services other than two-party telephones. We need to ask what the needs for travel are and what facilities, if any, could satisfy these needs.

In some cases persons travel to meetings. It may be desirable for many persons, rather than two, to confer. In this case some form of telecommunications conferencing facility may serve as a replacement. There have been experiments with voice (telephone) conferencing in which there is an arrangement for indicating who is speaking at the other end. Some users of telephone conferencing systems have reported that such meetings are more disciplined and businesslike than face-to-face meetings.[5] At the same time they are more tiring, because the participants have to concentrate more. There is less social chitchat and diversion, and this lessens the attractiveness of such meetings. Many conferees would rather meet face to face.

In some cases (but not always) more can be achieved if video links such as television are used between the conferring parties. In some cases, travelers feel that eyeball-to-eyeball confrontation is necessary. Would Picturephone or larger-screen video links then suffice? Or is there something more subtle in the psychology of communications that electronic media cannot convey?

There is now substantial experience of video links being used for business purposes in a smaller number of organizations. A typical reaction to the experience is that of the President of Dow Chemical commenting on the use of video links between Dow plants. He states that the facility is extremely valuable in making decisions that might otherwise not have been made and that it saves many hours in traveling and wear and tear on employees. At the same time it can never completely replace face-to-face discussions. There are some cases when the individuals *must* get together in the same location.[3]

One banking organization that is strenuously concerned with cutting costs is Citibank of New York. The organization has specially designed video-conference rooms in New York and a teleconferencing link to the First National Bank of Boston. Users of the system state that on certain occasions critical financial decisions have to be made quickly by people in distant buildings. The video-conferencing systems permit these decisions to be made satisfactorily, and that pays for the high cost of the links.[6]

A video-conferencing system has also been employed at Bell Laboratories. When it was assessed by its users, 91 percent said they would rather use it than travel fifty miles to a face-to-face meeting and 51 percent said they would rather use it than travel fifteen miles to a meeting.[7]

Several state and local governments in the Connecticut–New Jersey–New York area are connected by a video-conferencing system operating with privately installed line-of-sight radio links to and from a control center in the World Trade Center in Manhattan. The system is used for training courses and seminars as well as for meetings. Its users claim that the costs compare favorably with those of traveling to meetings.

The types of situation where electronic links seem inadequate are those where there is a substantial emotional involvement. Sales representatives will be reluctant to use video links for closing important deals. It is easier for a client to say "No" if the sales representative is not there in person. Personal presence is needed for tracking down a reluctant client or getting a foot in the door. In some highly contentious arguments, face-to-face resolution may be desirable. For visits to factories or tours of inspection, visits in person will still be necessary. The Institute for the Future conducted research on the psychological effects of teleconferencing and concluded that 85 percent of a traveler's requirements could be satisfied with facilities for video conferencing and fast facsimile transmission.[3]

Certainly, most of the *internal* meetings in my own corporation could be carried out effectively by video conferencing and facsimile. Perhaps more important, there would be much communication between people that is lacking today because the parties are geographically separate and cannot be constantly traveling. In many corporate situations this lack of continual communication does much harm.

If good video conference facilities *do* become available, some businesspeople will be slow to adapt to them. For instance, the British Post Office has used Contravision service, which provides public teleconference studios in several cities, and it found little criticism of the *effectiveness* of the service. However, British businesspeople were slow to use the new service. There appears to be an inherent conservativeness in how some people communicate, and British businesspeople were reluctant to give up the pleasures of travel and the personal contact, drinks, business lunches, and general socializing that went with it. There are strong cultural differences between nations, and some nations will more easily adopt new communications techniques than others.

COSTS

For most organizations the cost of video links is far too high to consider. Because there is no public video service they would have to own or lease their own links. At the time of writing, Banker's Trust of New York paid $11,000 annually for a 3½-mile link that was used only 1½ hours a day.[3] The Dow Chemical video link, spanning 1400 miles, averaged $2000 per hour of use.[3] The Connecticut–New Jersey–New York system, spanning distances up to 50 miles, cost $275,000 per year.

A large communications satellite could provide a public teleconferencing service in which the links are shared by many corporations and many types of users. NASA's ATS-6 satellite demonstrated the feasibility of this, using relatively low-cost earth antennas. The cost of long-distance video conferences could drop from $2000 per hour of use to $200 per hour or less, and much lower when there is widespread use of larger satellites.

Airfare savings paid for the Dow Chemical link. However, in addition Dow Chemical calculated that traveling for a meeting between its video-connected plants took a minimum of 15 work hours of an executive's time for a one-hour meeting. This time is too valuable to lose, and in addition most executives experienced much wear and tear from the trips.

The Dow System links its headquarters in Midland, Michigan, to its manufacturing division in Houston, Texas. *Mobile* color television equipment is used. Both ends of the link have meeting rooms with six-foot by eight-foot rear projection screens. The system is used for technical conferences, laboratory meetings, product business team meetings, and management planning discussions. Large corporations spend a great amount of time on such activities, and they are well suited to handling by video links. If appropriate satellite facilities were available, the patterns of business travel would change.

EXAMINATION OF DOCUMENTS

Many persons travel because they need to see and discuss documents, drawings, or other matters in more detail than can be accomplished by telephone. They are not particularly interested in face-to-face or emotional contact. To these persons the right telecommunications facilities not only could replace their trip but also would be much better than travel, because they could communicate repeatedly at any time and could communicate separately with persons in many locations when they were concerned about a particular topic.

The right communications medium for them is not Picturephone, or even television, because these cannot show enough detail on the screen. The transmission could be *much* less expensive than Picturephone or television, because *still* images and voice would be transmitted rather than *moving* images. The transmission capacity required could be less than one one-hundredth of that for today's Picturephone or television. Users could inspect documents together while talking, pointing to them on the screen, and transmitting a printed copy if desired. In some cases the screen unit may be a computer terminal.

There are many possible forms of communication at a distance. We will explain them further in the next chapter, which is concerned with the requirements of industry.

LOCALIZATION

A pattern of much more *localization* of physical activities may emerge in the future. Physical activities will be centered in local communities; information activities will be nationwide or worldwide via telecommunications. Today the bread a Frenchman eats has usually come from a nearby local bakery; the bread an American eats has often come from a thousand miles away. When transportation is cheap, the economies of scale of mass production and mass distribution can provide goods more cheaply. When transportation is expensive, as it was a century ago, local production is preferable. Mass distribution on a vast scale has become the way of life in America. One can look down from an airplane at a landscape of hundreds of small towns surrounded by fertile farmland, but the food sold in the supermarkets of those towns is not that grown locally; it is food packaged, advertised, and distributed on a nationwide scale.

Local communities in the future may grow more of their own food and provide their own daily needs. They will have offices for white-collar workers plugged into nationwide telecommunications networks. They will have satellite earth stations or other links that provide the same television facilities as in the big cities. The local bread and vegetables will be better than those that are mass-packaged for nationwide distribution. Much of the drudgery of commuting will be ended. Particularly important, the job opportunities in local communities will be almost as good as in the cities. There will be no need to uproot and go to the cities to become wealthier. For many the life style of rural communities with excellent telecommunications will be preferable to that in the cities.

VIRTUAL CITIES

In the past, large cities have had a major effect on human culture. The large accumulation of talented people in one place has a chain reaction effect. Eric Hoffer commented on cities as follows:

> Not a single human achievement was conceived or realized in the bracing atmosphere of steppes, forests or mountain tops. Everything was conceived and realized in the crowded, stinking little cities of Jerusalem, Athens, Florence, Shakespeare's London, Rembrandt's Amsterdam. The villages, the suburbs, are for the dropouts . . . we will decay, we will decline if we can't make our cities viable.

Today cities are in various forms of trouble. Some are economic disasters. Some have appalling levels of crime and violence. Some are foul with pollution. Many in the developing world are hopelessly overcrowded. A popular comment in Brazil is, "If you believed in the Twentieth Century, go and live in Sao Paulo."

The growth of a telecommunications society would relieve some of the population pressure on the cities. It could greatly reduce the pollution, raise the quality of life, and lessen the drudgery of commuting. However, it may worsen the financial problems of the cities, because corporations and individuals who pay the most taxes could more easily locate elsewhere, whereas the poor of the cities cannot.

Many authorities forecast the growth of a new rural society, and census figures already show a reversal of the trend of population movement from country to city in many industrial nations (the opposite trend in developing nations).

The growth of the city was culturally essential in Victorian times. Only in large cities could sufficient numbers of like-minded persons gather to foster new activities. Personal contacts had to be made face to face. Victorian London was spectacular for its gathering together of intellectuals, its new societies, its theater and music, its bankers and brokers, and its corporate associations. Like other great cities, it was throbbing with new ideas, new culture.

With telecommunications and jet travel, much larger and more specialized groups of authorities can be drawn together. Dramatic arts and music can flourish on television, radio, films, or recordings. Like-minded persons convene worldwide. We can ask to what extent Hoffer's comment will still be valid in a telecommunications society. Will excellence and creativity flourish on rural campuses and in global interest groups connected by video links and computer networks? Few can doubt that they will. Does the satellite age make the city obsolete? Probably not entirely. Whatever New York's problems, and however much of its business and population disperse, there will still be many who need the excitement of its streets, its anonymity, its bars and sex, its theater, its galleries, and the feeling of proximity to exciting people. New York will not be dead in the satellite age, but the role of rural communities will be greater enhanced. The functions of a city will be different.

Using the jargon of the electronics industry we could talk of "virtual cities"—communities, campuses, laboratories, or corporate offices, scattered across the earth but connected electronically so that the chain reaction of human stimulation catches fire as it did in Victorian London. Virtual cities will not make real cities obsolete. They will be different, just as movies are different from live entertainment. The physical presence of stars, of professors, of lovers, of ordinary people makes a vital difference. But the existence of virtual cities of innumerable types woven around the globe will enormously enrich human culture.

THE CHANGING GNP

America and other nations have changed from being predominantly a farming nation to predominantly a manufacturing nation. In the 1920s and even the 1930s most Americans considered agriculture the basis of America's economy. By the 1950s it was clear that manufacturing had become the paramount economic force. Social values and institutions changed in accordance with this perception. The government spent money on the infrastructure needed to make a manufacturing society efficient.

Recently, another change has been taking place—the growth of knowledge industries, which produce and distribute ideas and information rather than manufacturing goods. This accounted for only a small fraction of the gross national product in 1900. By 1955 it had grown to one quarter of the gross national product, and by 1965 to one third. Now one half of the gross national product is earned and spent in producing and distributing information and ideas. The percentage will continue to grow as manufacturing becomes more automated and as consumption patterns change under the influence of the forces described in this book and of the shortage of oil and raw materials.

This changing nature of work and consumption needs a different infrastructure to support it. Instead of highways for physical transportation we need highways for information. These new highways—video links, electronic mail channels, computer networks—are seriously inadequate today for society's needs. In the 1960s the U.S. Federal government alone spent $70 *billion* on road construction. To make the electronic society operate efficiently—to make the gross national product grow in the new era—similar massive expenditure on electronic highways is needed.

Chapter Seventeen

INDUSTRY

The cost *of national mismanagement of telecommunications has been very high in some countries. It will be higher in the future.*

Much of what we discuss in this book relates to the *social* impact of telecommunications; however, possibly the most important impact of telecommunications will be on *industry*. The wealth of advanced societies is generated mainly by industry. Social improvements can occur only if sufficient money is available. A society or community is in deep trouble if its expenditures continually exceed its wealth generation. There have been a few examples of this in the 1970s. New telecommunications facilities can have a major effect on the efficiency of industry and hence on the wealth of nations.

Means of communication or transportation has always been an essential ingredient of economic growth. Those nations which have had the best facilities for distribution have tended to be those which rose fastest to economic power. The first industrial revolution was able to sustain its mushroom-like growth because transportation facilities were built, canals were dug, and weary horses spent their lives lugging barges of coal from the new mines to the factories. Many factories clustered round the coal fields, and whole segments of the population moved there into gloomy, jerry-built rows of houses. It took many years for the means of transportation to become adequate for the needs of the new industry. Later the railways were built, and after electricity was invented overhead wires carried the necessary power across hill and dale. Britain built the largest empire in history because it excelled in the means of overseas transportation, its navy. America became the world's richest nation after it built great railroads and, later, the finest highway network and a trucking industry more efficient than any other country's.

Communication links are a vital part of society's infrastructure, and in a society where more than half the work force works with information and ideas, we need *electronic* communications. Manufacturing will be increasingly run by process-control computers and production robots. Paper work will be increasingly handled by data processing. Management will have an insatiable desire for information, and those with the best information will succeed over their competition. Vast data bases will be built up and will serve many locations. Computer networks will lace together the corporate and government facilities. Small computers will become as common as adding machines, but many personnel will need to use the million-dollar systems and data bases, which may be a distant, shared resource.

Speed is often important in the use of computers. A managing director wants a *quick* answer to a question in order to make a decision. A doctor wants a *quick* analysis of electrocardiac or other data. A factory shop floor must be controlled in "*real time*" as its events are taking place. There is not enough time to write a letter to the computer center.

As in the past, the nations that have the best communication facilities may have the highest economic growth. The corporations that have the best electronic communications may surpass their competition. "He who runs the information runs the show."

Most corporations today spend a substantial proportion of their money and human talent on communication of one form or another. Businesspeople spend most of their day communicating. Most white-collar workers spend much of their time communicating with superiors, subordinates, customers, suppliers, secretaries, and computers. *The information-handling process typically costs from 5 percent to 30 percent of an organization's total expenses.*[1] Given such a large expenditure, it is desirable to ask how corporate communications can be made as effective and inexpensive as possible.

During the industrial revolution the means for distribution lagged seriously behind the need for it. Today there is such a lag with the new machines. In the first decade of widely accepted commercial use of computers, there was hardly any data transmission, except in a handful of pioneering experimental installations. Now, data transmission has become vital, but there are problems. The telecommunications links and switching were mainly designed for handling telephone conversations, not computer data. Equipment that could send and receive electronic mail has long been available, but we spend a fortune on secretaries, mailrooms, mail clerks, and the Postal Service, partly because our telecommunications channels are not quite suitable for electronic mail. Television equipment has been around for decades; yet corporations spend massive sums on airfares and other physical travel because telecommunications links suitable for video conferencing do not yet exist.

The electronic links needed for running industry and government efficiently are fundamentally different from the public telephone network. The *technology* for building these links at the right price now exists.

COMMUNICATIONS COSTS

If we look at telephone communication alone, there is much scope for lowering costs. Table 17.1 shows typical U.S. costs for traditional corporate telecommunications, most of which is telephone traffic.[1] Large corporations spend many millions of dollars on their telephone facilities.

Table 17.1 Cost of corporate telecommunications as a percentage of total operating expenses.

Industry	*Range (percent)*	*Average (percent)*
Airlines	3–7	4
Banking and finance	0.6–4.2	1.5
Insurance	1–3	2
Manufacturing	0.3–2	0.5
Securities	8–12	10

Surprisingly, about three fourths of this expenditure relates to internal telecommunications. A widespread corporation can therefore profit by having an internal telephone network designed to minimize costs. In spite of the magnitude of the costs, there is often no technical study of corporate telecommunications or high-management involvement, as there would be with other equipment of similar cost. Accountants take most communications expenditure for granted and do not try to justify its cost as they would with, say, computing equipment.

In addition to telephone networks, corporations have

networks for transmitting data. The volume of computer data transmitted in many corporations is growing at more than 25 percent per year and is increasing by a factor of 10 between 1970 and 1980. There are often multiple data networks forming parts of different computer systems that were designed separately.

Today, corporate telecommunications is generally thought of as telephone and computer traffic, possibly with a few small extras such as occasional facsimile messages. The total cost of communicating in a corporation, however, is much greater than the cost of telephone and data transmission—typically between 5 and 10 times as much. Two major contributors to this cost are the sending of mail in a corporation, especially internal mail, and the cost of physical travel for communication.

An average price of correspondence in the United States has been estimated to cost $10 or more to be conceived, formulated, copied, transported, received, read, and filed.[2] This is much higher than the cost of making a telephone call. A very large quantity of memoranda and letters are sent within most corporations. Modern telecommunications enables us to ask a new question: Should not corporate correspondence be sent in an electronic form rather than in the form of paperwork? Electronic memoranda have two important advantages. First, given appropriate systems design they can be cheaper. Second, they can reach their destination faster. The highest-priority correspondence can reach its destination in seconds, so that two people conversing by telephone could exchange a document while they talk.

ELECTRONIC MEMORANDA

There are a variety of ways in which memoranda can be handled electronically:

1. *Telegraphic message switching.* Memoranda are typed into terminals like telegraph machines and delivered by a store-and-forward system. They may be filed by the system, not by filing clerks.

2. *Visual display message switching.* Memoranda are typed into screen units that permit easy editing of documents. Retrieval on screen units avoids paper handling. Screen units on executives' desks would be used for many other functions.
3. *Magnetic card systems.* Magnetic card typewriters permit easy editing and storage of documents and may form the input to a message-switching system.
4. *Facsimile.* Facsimile machines permit drawings, signatures, logos, and handwritten notes to be sent. If transmitted or stored digitally, facsimile documents require about ten times as many bits as alphanumeric documents. In some cases use of facsimile can save typing costs.
5. *Voice message storage.* As electronics costs drop and secretarial costs rise, the cheapest way to transmit and store interoffice messages may be in the form of speech. Speech messages could go from the sender to the receiver, like telegraphic messages, and avoid any intermediate human processing.

SPEECH MEMORANDA

Executives spend much of their time dictating messages, which secretaries must type, to other people in the corporation. The most labor-saving and potentially cost-saving form in which to deliver such messages is spoken-voice form. The spoken-voice messages (viocegrams) would be filed in computer storage. Each recipient would be notified when messages were waiting, or one could check one's file periodically from an ordinary telephone. The cost could be lower than the cost of sending and filing typed memoranda.[3]

The use of speech memoranda instead of typed memoranda would be alien to some executives who cherish their present way of operating. However, it would be more convenient to send many memos by telephone. Whenever an executive cannot reach a person by telephone, he or she can immediately leave a message with no danger of secretaries misinterpreting it. When traveling, an executive can telephone the system, at night if necessary, and listen to memos that have been stored for him or her. Secruity controls can be designed to prevent any other person listening to the messages. Such a system could be more convenient and accessible than a memo-typing operation.

Memos are used in corporations to provide a record of past instructions and communications, and the memo sender signs the document. The voiceprint of an individual saying certain words, for example speaking his personnel number, is as good an identification as a signature. The system may be designed to store speech memoranda in long-term files when instructed to do so. When not so instructed, it keeps them for a given period of time after they are delivered (perhaps a month) and then erases them.

Memoranda contain tables, diagrams, or other items not conveniently represented by speech would still be sent on paper. The paper may be delivered by facsimile or alphanumeric coding and be stored in the same system as the speech memoranda. Speech memoranda would not be appropriate when files have to be searched.

CORPORATE TRAVEL

Not only is the total cost of typing and filing memos greater than the telephone bill in most corporations, but the total cost of business travel is also greater—and the cost is rising. Some large corporations spend more than $100 million per year on business travel within the United States. This cost does not include the time of the persons traveling or the effects of wear and tear on them, which for some executives are considerable.

As we discussed, telecommunications can form a substitute for certain types of travel, with massive savings in cost and human time. It is necessary to ask the question: What are the best ways for people to communicate with people at a distance? This is a complex question, and there are many types of answers to it. Many of the answers require transmission at higher rates than that of local telephone loops, often in very brief bursts.

The following are ways in which corporations could find substitutes for travel.

1. *Telephone conference calls.* Conference calls have been infrequently used in business because of the difficulty of setting them up. Some new computerized telephone exchanges give their users the ability to set up multiple-party voice connections without operator intervention. This can be a valuable facility, because it is often very useful to consult a third party during a conversation or to include several people in a telephone discussion.

   A telephone "meeting" with many parties at separate locations needs a certain discipline imposed on it to make it effective. It is necessary for each person to know who is speaking and to be able to indicate to the dispersed group that he wants to speak. This can be done if the meeting has a chairperson who disciplines the conversation. Another possibility is for each caller to have a small strip with lights. One of the lights is associated with each participant; it is on when he is talking, and he can make it flash when he wants to talk. A low-bit-rate control channel is derived from the speech channel for operating the lights. The strip may be designed so that a caller can write the names of the parties by the lights assigned to them.

   Telephone meetings may also be held between meeting rooms in different locations equipped with phones that have loudspeakers.
2. *Picturephone.* As we discussed in Chapter 15, Picturephone adds information to a telephone call by permitting the observation of facial expressions. The cost of this extra information is high—many times the cost of a telephone call—and for many corporate calls it is not worth it.
3. *Facsimile.* Paper documents can be transmitted fairly quickly by facsimile means. Telephone callers may use facsimile transmission to enhance their conversation, so that they can exchange sketches or documents and discuss them.
4. *Still video.* In conjunction with a telephone call, callers may employ a screen on which still images can be displayed. If the image is to be displayed and dis-

cussed while the conversation proceeds, it is desirable for it to be transmitted fairly quickly—say, in five seconds. If the telephone call is occupying 64,000 bits per second on a digital channel, then the image could go over the same channel, interrupting the speech in one direction for five seconds.

To enhance the conversation, both parties should be able to look at the same image at the same time and point to it. A movable arrow may be provided on the screen for this purpose, along with a low-speed subchannel for conveying its movements.

It may be desirable to see the caller's face, but as a still image rather than the moving image of Picturephone, which takes so much transmission capacity. The lens used to transmit documents could also be used, as a Picturephone lens is, to record the face of the caller. A system could be designed in which a telephoner could press a button to capture the image of the face he is talking to. He could then have some idea of a person's expression at a selected critical moment in the conversation. This would not convey as much human information as Picturephone, but the image could be transmitted over a telephone channel.

5. *Electronic flip-charts.* Most major locations in corporations will have visual display units connected to computers. Such screens have been used effectively for enhancing person-to-person communications as well as person-to-machine communication. Two or more individuals talk by telephone and discuss information that resides in a computer storage. All participants to the conversation see the same data displayed, and all can modify it.

   At its simplest level the machine is being used merely to display human ideas with clarity. In many corporations today flip-charts are used for this purpose. An employee has a set of facts or ideas that he must present quickly and efficiently to management or colleagues. He writes the information in a concise form, so that it will be grasped quickly, on large sheets of paper which can be hung on a stand. Employees travel with a roll of flip-charts to make half-hour presentations to management. The information could be conveyed equally well if the data on the flip-charts were entered into a computer system as a set of single-screen displays and edited until they were as concise and clear as possible, preferably in color. The persons talk by telephone, using the electronic flip-charts in the same way as paper flip-charts.

Electronic flip-charts have several advantages other than avoiding the need to travel. First, they remain in the computer storage after the conversation. Management personnel rarely admit it, but they often remember only a portion of the data that is flip-charted at them. It would be useful if they could review the charts again privately, at their leisure, and perhaps discuss them with persons other than the original presenter.

The production of paper flip-charts is often made a laborious task. Neat lettering with a felt-tip pen takes time, and the wording is frequently modified. Computer software could make the entering and editing of screen charts a fast operation.

In a corporation the many flip-chart presentations could be filed and indexed, with appropriate security locks. Many flip-chart presentations are made on the same or related subjects, and the indexes would permit computer searches to be made for these.

In some corporations flip-charts are one of the main forms of communication, with much money being spent on air travel by persons making flip-chart presentations. It can be a highly efficient form of communication and is susceptible to mechanization. Computer-assisted flip-charting has major advantages.

6. *Communication via a data base.* Communication links that handle flip-charts could also handle data that are assimilated and stored by computer data base systems. Data base technology imposes a measure of precision on the way data are defined and referred to, and data base administrators have often been surprised by how different departments or managers call the same data by different names or different data by the same name. When communication takes place between parties using a common data base, there is less chance of imprecision.

   In a system at Westinghouse, a graphics terminal

is used for production scheduling based on sales forecasts of washing machines. This is a complex operation, because Westinghouse makes over one hundred models, all available in several colors. Once a month the production and marketing managers travel to Pittsburgh to work together on the display console. The marketing managers evaluate market forecasts and assist the production managers in working out the production schedule. The use of the terminal permits more options to be explored than were possible before. Before, a "seat of the pants" approach was necessary; now the two groups of managers can communicate with precision. The managers involved, once experienced with the technique, "wouldn't want to do their scheduling any other way."[4]

The same type of meeting could take place via telecommunications links, with the parties involved able to discuss data that all can see. In some cases the data will be modified or processed during the conversation.

In reflecting on the ideal form of person-to-computer dialogue, we find that they have much in common with ideal forms of person-to-person dialogue via machines. As person-to-machine communication improves and person-to-person communication becomes more precise, the two will increasingly tend to require the same hardware, channels, and features.

7. *Teleconference rooms.* Teleconference rooms for holding meetings may become a major corporate facility.
8. *Radio paging.* Radio paging systems make it possible to contact individuals who are not sitting near a telephone. The individuals wear a small, inconspicuous radio receiver that can signal them either with an alarm tone or with a spoken message. Individuals roaming about a factory floor can be instructed to pick up the nearest telephone. Service personnel miles from anywhere can be instructed to go to a particular customer. Paging is one of the facilities of some computerized telephone exchanges, but it is more often done manually. A few corporations make massive use of radio paging.
9. *Two-way mobile radio.* Two-way radio can be used in the ways discussed in Chapter 11, and it has a major

growth ahead. It can be used to control delivery or other vehicle fleets in a way that reduces the total distance traveled.

Two-way communications to persons far from conventional telecommunications facilities is being accomplished via satellite. Satellite links connect to off-shore oil-drilling rigs. The MARISAT satellites connect ships around the world to their head offices. Until now ships have been million-dollar facilities without the transmission capability of a terrestrial office. Now they can be linked into their corporate communications network, like any branch office.

Remote facilities on land can also be linked to corporate networks by radio, possibly via satellite. For example, the trucks that service tractors or earth-moving equipment in developing countries can have radio links to their offices.

WORLDWIDE NETWORKS

The world's largest private telephone network is that used by the U.S. military—AUTOVON (AUTOmatic VOice Network). Its total circuit mileage in 1970 was equivalent to the entire Bell System in the early 1950s. It handles data as well as voice communications, and calls can be encrypted for security. The circuits link U.S. military installation all over the world. In the future, corporations will have worldwide networks that incorporate many features of AUTOVON. They will be a fraction of the cost of AUTOVON, because of new satellite facilities and because they do not need to be nuclear-bomb-proof. A corporate network needs reliability, not "survivability." Like AUTOVON, it should handle data and voice over the same lines. It should also handle electronic mail and messages. As on AUTOVON, some calls will be highly important and should be connected as quickly as possible, but the majority of calls will be low-priority administrative traffic. It is desirable for certain top executives and their assistants to be able to contact each other immediately. On AUTOVON all "command and control" calls are set up quickly, and this step could be profitably copied in industry.

Today executives commonly receive busy signals from the tie-line network because the lines are flooded with unimportant chitchat.

AUTOVON uses pushbutton telephones with four red keys to the right of the normal twelve keys. Pressing the "P" key designates the call as "priority," in which case it preempts ordinary calls. The "I" (immediate) key preempts priority calls. "F" (flash) and "FO" (flash override) give the highest levels of precedence. When a general presses "FO," his call blasts its way through all other calls. The called party receives a special "precedence" ringing signal. A unique tone informs telephone users when they are being interrupted by a higher-precedence call. If one of these keys is used on a telephone not authorized to use that level of precedence, a prerecorded voice tells the caller that such a call cannot be put through. Top executives ought to have the same facilities.

In addition to the ability to make precedence calls, the general can have one or more "hot lines." He picks up a red telephone on his desk and is almost immediately connected to a predetermined location. Alternatively, he can press one key on a telephone with the same effect. The top executive in industry might have a "hot line" to a personal assistant, distant plant managers, or presidents of subsidiary companies. In the future he or she is quite likely to have a hot line to an industrial "war room." The hot-line connections on AUTOVON are, in fact, switched. This factor makes them lower in cost and more reliable. A priority level is automatically assigned to such calls, and service is so fast that a user often does not know that the call is switched. By the time the handset reaches the ear, the telephone of the hot line appears to be ringing.

Another function of AUTOVON is automatic setting up of conferences. A number representing a preselected list of conferees is keyed; the system looks up their numbers in its memory and switched lines to them. The conferees in this automatic hookup can be located all over the world.

In future private networks, a computerized exchange may be programmed to search for an individual, to try a number repeatedly if it is busy, or to page someone by radio.

The extension of the hot lines by radio paging could make key individuals permanently accessible.

WAR ROOMS

Since the earliest days of the computer, a trend from military usage of information systems to industrial usage can be detected. In the beginning SAGE, a vast system in the United States for detecting and tracking hostile aircraft, pioneered real-time processing, visual display screens, and light pens; ten years later these devices were common in industrial systems. Now the technology of military command-and-control systems is being found in industry. Data transmission networks are spanning commercial organizations. The techniques employed in "war games" are finding uses in industrial and civil situations.

A term commonly used in industry now is "war room." A war room is a communications nerve center from which many information sources are accessible. Critical corporate decisions may be made with the aid of the war room.

Obtaining information from the various computers has become a highly complex process in a large corporation. Many systems do not attempt 100 percent automation but use a judicious mixture of computers and human experience. One way to do so is to gather together the talent and terminals in rooms equipped to answer or redirect management's questions. Such rooms exist in embryo form today in a variety of organizations, and it is clear that they will grow in complexity and diversity of functions. They range from the war rooms of military command and control systems to centers for helping eliminate errors in data collection systems. In controlling a complex set of operations, the decisions that must be made are often brought together by telecommunications into one nerve center. A spectacular example is the NASA control center for the space missions at Houston, Texas. Many less spectacular examples exist in industry.

The showpiece of many data processing installations in the future may be a war room in which all types of information can be sought from different remote computers and routed to locations where needed by management, sales staff, shop foremen, and others. The computers and their storage, no longer glamorous showpieces, will be hidden in secure locations.

The manager of the future may have a hot line to the local information room. He or she may pick up a red telephone on his desk and be immediately switched to personnel who know where to find the answers to various types of questions. If possible, they will answer his question verbally, or display relevant information on the screen in the manager's office, or prepare a printout. Sometimes they will route the query to another information room, perhaps more specialized, perhaps more expert, or perhaps in a distant part of the organization.

A VITAL INFRASTRUCTURE

New technology, including data networks and satellites designed specifically for corporate use such as those planned by Satellite Business Systems (a subsidiary of IBM, Comsat and Aetna),[5] will form a part of a nation's infrastructure that is vital to its industrial productivity. Some nations will have these new communications channels in the near future; others will not. Industry in those countries which have them will have a competitive edge over that in countries without them.

The *cost* of national mismanagement of telecommunications has been very high in some countries. It will be higher in the future.

Chapter Eighteen

3½-DAY WEEK

An individual is systematically manipulated to fit into a system that evolved prior to the electronics revolution. The prospect of electronics is an end to human drudgery.

The prospect that electronics offers us is an end to most human drudgery. Boring jobs can be done by computers communicating with one another over data networks. Factories can be largely automated, each machine tool containing a microprocessor, and the overall operations being controlled from telecommunications nerve centers. Much of the boredom of commuting can be avoided by geographic redistribution made possible by new communicating links. Tedious jobs like typing memos and mail delivery can be largely avoided. Bank tellers are no longer necessary, and electronic fund transfer can replace the enormous tedium of processing checks and bills and credit card paper work.

While the machines of a wired society telecommunicate, humans can devote themselves to more interesting pastimes.

The bank teller and the letter carrier, however, do not necessarily share this euphoric view. For them and for persons in many other routine jobs, the prospect of unemployment arouses fear. Workers band together in unions to fight the spread of technology that reduces jobs. Unemployment in the 1970s is higher than it has been for decades, and it remains high during years of economic prosperity. There is concern that it may never sink to the levels of the 1950s or 1960s. Tedious jobs would be better than no jobs at all. Unemployment is particularly serious among young people, including college graduates, who are not yet an established part of the work force. The unemployment rate among the young is often three times the national average rate.

As automation gains more muscle with intelligent machines, this argument will become more intense. We will be increasingly faced with the following choice. Can we find more work for people in spite of automation, or should we restructure work patterns so that people work less? An alternative to a high unemployment rate is a reshuffling of work so that everyone works shorter hours.

The machinery used by society is becoming steadily more expensive. Automated factories, thorough pollution control, and advanced communication facilities require much capital. The equipment requested by research physicists and space scientists is becoming so expensive that often it can be bought only by a consortium of nations. Hospital equipment is becoming extremely costly. This trend to even more expensive facilities is likely to continue.

When large capital expenditures have been made on equipment, it is desirable to keep it working full time. This will help maximize the financial return on the investment. It will be increasingly desirable to keep the automated factories and giant computer systems running seven days a week. In order to do this in a society characterized by high automation and a desire for leisure time, it would make sense to have a 3½-day work week. Persons could work three days one week and four days the next, alternating three- and four-day weekends. The machinery and computers would work a seven-day week, run by alternating staff. We should slave-drive the machines in order that *we* can have more leisure, or a richer life.

Some persons might prefer to work three-day weeks. In this case three-day staff and four-day staff might be interleaved. Other persons might prefer to have seven days of work alternating with seven days of leisure. There are many ways of dividing people's working time that would keep the machinery going full time. In the computer age, complex scheduling of working patterns is practical. The computers can interleave different working patterns of different people so as to achieve full (seven-day week) staffing.

Some persons will object to working on Sundays or religious holidays. The scheduling of other people would be designed to compensate for this, possibly using higher pay on Sundays as an incentive.

Economists often object to the argument that increased leisure is a substitute for unemployment. They claim that it would lower the gross national product and swing the nation into a severe recession. However, if increased leisure could be introduced in a way that increases the utilization of productive capacity, that argument is not necessarily valid. Furthermore, many people would choose to carry out hobby activities during their time off, such as growing vegetables, hobby computing, running a motel, writing, developing teaching programs, running their own business, and so on. Some of these vocational activities would add to the gross national product.

We look back with horror at the work patterns of the industrial revolution, when children worked twelve hours a day in the mines. After the electronic revolution has matured, people will look back in horror at *our* working patterns: Five days of boring, mandatory work interspersed with weekends that are too short. Lengthy commuting into cities. Executives returning home too late to play with their children, too shattered to do anything but drink and watch television. Many people feeling that their jobs are meaningless—that the worthwhile part could be done by machines and the part that cannot be done by machines is valueless politics.

At the height of New York's economic crisis there were countless thousands of city government office workers. They have insufferably boring jobs, spend much of their working day creating subconscious diversions, pretend that their politics and bureaucratic rules have value, and relentlessly resist attempts at automation for fear of losing their entrenched routines. Most of their activity, when analyzed in enough depth, is unnecessary; but bureaucracy protects its own procedures so fiercely that New York had to fire police officers, fire-fighters and hospital workers rather than the vast office staffs.

Most of today's young people look in horror at society's work patterns. Work and the life patterns of their parents seem pointless and empty. There are better things to do with life. There is no shortage of worthwhile projects and fascinating hobbies, but the available work lacks meaning: making and selling useless products or serving bureaucratic structures. For most young people work is mindless, servile, exhausting, something to be endured; life is short and wonderful, but five sevenths of it is thrown away.

Beginning with school, or before, an individual is systematically manipulated to fit into a system that evolved prior to the electronic revolution. The child is forced into rigid patterns which suppress the imagination, creativity, dreams, empathy with nature, and personal uniqueness. The electronic revolution will not do away with work, but it does hold out some promises: *Most boring jobs can be done by machines; lengthy commuting can be avoided; we can have enough leisure to follow interesting pursuits outside our work; environmental destruction can be avoided; and the opportunities for personal creativity will be unlimited.*

The increase in possibilities for electronic entertainment and education go hand in hand with the potential for increased leisure.

NOT FOR EVERYONE

It is certainly not the case that *everyone* is likely to work a 3½-day week. Scientists, writers, film makers, and most creative people will burn the midnight oil and keep hard at their projects, as they do today. Executives entangled in corporate politics will probably not dare to leave for 3½ days. It is on dull or laborious jobs that time off will be welcome.

The telecommunications society will need more creative people than today, and this raises the prospect that while part of humanity will be freed to enjoy more leisure, part will work sixty-or seventy-hour weeks. Some will be running worldwide activities via telecommunications; some will be creating the television needed to fill the numerous channels. Others will spend their time sitting in front of their wall screens with a carton of beer.

It is difficult to predict all the social effects of a substantial increase in leisure, just as it was difficult to predict the social effects of the automobile. The rich classes of the eighteenth and nineteenth centuries spent their time in many different ways. Some were bored and decadent. Some gambled to excess. Some worked with furious intensity to increase their wealth even further or to build industrial empires. Some built extraordinarily beautiful homes or landscaped the countryside with lakes and woods.

A small proportion of the future leisured class will probably be far more creative than they could possibly have been with a full-time job. Victorian England had more creative people than any previous era in English history. It was a society in which many people had become wealthy enough not to have to work hard, as the masses did. From the wealthier class came inventors, writers, amateur scientists, musicians, Darwin, the Bloomsbury group, Gilbert and Sullivan, . . .

In the electronic society there will be a great need for creativity. The dropouts in front of the television screens will need more programs to watch. Electronic educational media will be powerful but will need vast quantities of program preparation. Computers, more common than automobiles, will be programmable for an infinity of different applications.

USE OF LEISURE

A substantial increase in leisure will cause problems associated with how people spend their spare time. Society's more interesting recreational facilities are already overtaxed. America's national parks are blocked with cars. The beaches of Europe in summer are crammed with deck chairs and bodies. Many are foul with pollution. Rising tourism and rising population will strain these facilities much further. If this is combined with a half-time work week, the beaches and national parks will become intolerable to many. Many of the waterfalls, mountain tops, and beauty spots of the world will become cluttered with tourist paraphernalia.

Many tourists will head for the beautiful old towns and works of art. Here the situation will become even more frustrating. Venice and Florence can hardly double their present number of tourists, let alone take ten times the number. There is a limit to the number of people who can see—and enjoy seeing—Rembrandt's *Nightwatch* or the *Mona Lisa.* The Secretary General of the Council of Europe, L. Toncic-Soring, stated, "Tourism not only feeds on our cultural heritage, it is likely to destroy it utterly." The beauties of the past, limited in number, cannot be made physically available to a *high* proportion of the population.

The satellite age must create its own cultural heritage and leisure facilities. Affluence and leisure will put a greater strain on the old facilities than will the rising world population. The electronic culture will make possible warehouses of great movies, theaters with 360-degree screens, planetariums, centers for electronically sharing the culture of other nations.

It is possible that travel may decline rather than increase. Society may become decentralized with more small communities, small work locations, small farmsteads, and vegetable gardens, but with superb transmission facilities interconnecting these locations. Leisure in small communities will encompass country pursuits as well as the educational and cultural facilities that were previously available only in the big cities.

DIVERSITY

More leisure will probably bring more diversity to society. People will be able to pursue their own interests and their own forms of creativity whatever they may be, rather than be trapped in five-day-week drudgery. Some will work themselves to death; some will go fishing. Some will be creative; some will merely sit and watch. But there will be more worth watching.

WHY IT MAY NOT HAPPEN

Most young people, whatever their politics, welcome the notion of electronics bringing more leisure time. It is inevitable, *eventually,* that this process will change society as drastically as did the industrial revolution. There are several forces, however, which may prevent it from happening quickly. The rising generation should note these forces and regard them as enemies of *the good life.*

First and worst is bureaucracy. As automation takes over more activities, the bureaucrats of the world fight back with more complex procedures, more forms to fill in, more committees, task forces, and study groups, more government and more statistics collecting. The rules become more complex and are applied without thinking or common sense. Bureaucracy can find employment for the entire cross section of humanity, ranging from the most stupid to the most intelligent. The more intelligent people can be involved in bureaucratic "research." The term "research" once implied striving toward new discoveries and new ideas. It was done mainly by scientists. In the world of the bureaucrat it is done by statisticians, lawyers, sociologists, and so on, and produces lengthy reports that contain no new ideas and invariably conclude that more research is needed.

It will be a sad end to human inventiveness if we all work five days a week in government departments or corporations that have long ceased to be entrepreneurial. But that is clearly one way in which new and future electronic capabilities could be utilized. Telecommunications make possible bigger and more intractable bureaucracies. We should recognize this as one of the serious dangers of our age.

In a 5-year study of the British health service Dr. Gammon referred to "black holes" of bureaucracy, noting that a 51 percent increase in the administrative and clerical staff had been accompanied by an 11 percent decrease in the number of beds occupied. He wrote "In a bureaucratic system *increase in expenditure* will be matched by *fall in production* . . . Such systems will act rather like 'black holes' in the economic universe, simultaneously sucking in resources, and shrinking in terms of 'emitted' production." Useless work tends to drive out useful work. In a *small* company people pursue their self-interest by achieving as much as possible with a small number of people. In a bureaucracy or noncompetitive corporation people pursue their self-interest by expanding their empire, devising procedures that occupy more people, preserving their own departments, and playing political games which consume work.

Trade unions often oppose automation. On the other hand, unions have also made demands for shorter work weeks. It would make sense for unions to accept the fact that automation could be used to shorten the work week. A shorter work week could benefit the union members and give the union more members. Nevertheless, some unions will probably try to prevent the benefits we have described.

Another enemy of leisure is the skill with which Western corporations advertise and sell unnecessary products. It is unnecessary, for example, and very expensive to change the shape of automobiles every year. Many products are designed with planned obsolescence, which ensures a replacement market in the near future. Sometimes replacement is assured by planning to bring out and advertise a succession of new models. Sometimes it is assured by designing items to fail. For example, some electronic wrist watches are marketed with the electrical contacts in them unsoldered. This overcomes the extremely high reliability of their solid-state circuitry! Some fail as soon as the battery is replaced, but not within the three-month guarantee period. The marketing of undesirable goods raises the gross national product but is harmful to society as a whole.

Another reason why long weekends may remain a dream is simply inertia. Large organizations may not permit such a radical change. Employees may not want it, because they have always worked five-day weeks. Politicians will promise to lessen unemployment by creating *unnecessary* jobs. Homemakers may not want their spouses at home 3½ days a week.

Many people over 40 claim that they do not want more leisure time, or that increased leisure for the community would be bad because people would not know what to do with it. These are often people set in their ways who have long since had the innate creativity of human beings suppressed into the rigid channels necessary in a world of drudgery without automation. I can find few people under 21 who want to spend the rest of their lives working 5 days a week at a boring job from 9 A.M. to 5 P.M. On the other hand, youth are often frightened of the prospect of unemployment which is so high among today's young people. A 3½-day week would be a much better alternative to the 25 percent unemployment for the young, especially those who feel they have better things to do with life than spend it all at a boring desk job.

New technology gives us the riches to build a better world. Often we fail to take advantage of it because of inertia. It is important that today's youth should understand the potentials, and rebel against drudgery and bureaucracy.

Chapter Nineteen

EDUCATION

The most successful teachers in a telecommunications society will earn as much as film stars.

In a pioneering experiment in Brentwood School in East Palo Alto, California, one hundred small children entering the first grade were introduced to computer terminals with screens and light pens. The machines were used to assist in teaching reading and arithmetic. Brentwood was an interesting choice as the first school to use computer-assisted instruction, because it is in a slum area and had an 80 percent black enrollment and because the average IQ of the children participating in the reading experiment was measured at 89.[1] The technique was very effective. The children loved playing with the terminals, and their teachers had to "peel them off the machines" to get them back to their other lessons.

At the same time IBM was setting up a network to provide computer assisted instruction to its array of maintenance engineers. These engineers have periods of inactivity when no machines happen to need maintenance. They have a complex job and always have the potential of learning to work with more elaborate machines. The engineers around the world were given access to terminals that would teach them new technology in their inactive moments. The scheme became a successful and important part of IBM's operation.

Telecommunications can deliver education in a variety of new ways, either with or without the help of computers. This has two benefits of great importance to society. First, the skills of the few, rare educators of superb quality can be made available to millions. This is an amplification factor that will greatly increase the levels of education in society. Second, small communities and areas far from the sophisticated centers of learning can be given access to excellent instruction. The village one-room schoolhouse could provide high quality education, because it will combine the love and attention of a dedicated teacher with telecommunications access to the best instruction in the world.

Another interesting implication relates to teachers' salaries. Today teachers' pay is poor, and there is little incentive to excellence. In the future teachers who acquire a reputation will be wanted for the television links. Any teacher can attempt to make video tapes or write computer-assisted instruction programs. The organizations that distribute or broadcast such products will be on the lookout for talent;

their products will be watched by millions. The most successful teachers in a telecommunications society will probably earn as much as film stars earn today.

LIFELONG EDUCATION

Education used to be regarded as something that ended when one's working life began. In the electronic era it will go throughout life; *adult* education is of vital importance. The decades ahead will be characterized by an extremely rapid rate of change, in which both work and leisure activities will change. Many persons will learn two, three, or four careers in a lifetime as telecommunications, automation and, later, machine intelligence will cause entirely different work patterns. Electronics will create both the need and the tools for lifelong learning.

Work will become more complex, because most simple jobs will be done by machines. The techniques used by management and professionals will be more sophisticated and powerful. There is a need for much deeper and better training in government and industry.

A commonly held view is that it is difficult to learn new ideas when one is older. The main reason for this is that most people's brains become out of practice at learning at an early age. Just as muscles become weak if not used, so the brain becomes weak at learning if not exercised. Persons, often university professors, who spend their lives being perpetual students have no difficulty learning when they are old. In fact they sometimes learn more rapidly and thoroughly, because their brains are so practiced at learning and relate new ideas to numerous earlier ones, perceiving the similarities and differences.

Access to data banks and computers will greatly change what it is necessary for engineers, doctors, and other persons to know. A person well educated to practice a profession need not be a storehouse of facts and techniques. He can obtain the facts he needs from computer terminals and can instruct the machines to carry out tasks for which he previously needed lengthy training. A computer at M.I.T., for example, has been programmed to carry out skilled mathematics. It can solve simultaneous equations, differentiate, evaluate the most complex integrals, and perform most types of algebraic and calculus manipulation. There will be no need in the future for scientists or mathematicians to struggle with these tasks.

Schoolchildren already have pocket calculators. Before long they will have access to machines that do most of the unpleasant things in mathematics. Education will not need to stress how to solve equations or integrals other than simple ones. Instead it can concentrate on *how to use* mathematics—how to formulate problems so that the machines can solve them—which is the truly creative part of mathematics. The range of problems an individual can solve will then be immensely greater.

Education in technology is undergoing a transition. In many universities it traditionally used to emphasize know-how; today there is more emphasis on science and the development of mental agility, which will produce graduates who continue to grow professionally throughout their lifetime. This will make them better able to survive the increased complexity and future shock and better able to learn how to make use of new communications facilities and rapidly changing technology.

Doctors, like engineers, will be increasingly supported by electronics; this will make it less necessary for them to carry innumerable medical details in their memories, since these will be readily available on their telecommunications screens. This will make medical education less concerned with details and more concerned with principles. There are two general classes of medical education today: schools that teach much detail and produce physicians who can treat patients soon after graduation, and schools that teach principles and require internship and residency before their graduates are let loose on patients. The latter schools generally produce doctors who are better equipped later in life because they continue learning. The same will be true in many other professions, as electronic facilities increase both the complexity and the help available via telecommunications.

TELEVISION

Television has been used in education with widely varying levels of success. At one extreme, students have rejected it as impersonal and boring. At the other extreme, it can make education vivid in a way that no classroom lecture can. Alistair Cooke's programs on *America* teach history in a uniquely revealing and memorable fashion. The difference between success and failure lies in the quality of instruction, and that ranges over an extremely wide scale. To make programs like Alistair Cooke's *America* needs a unique personality, the highest level of professionalism, and a vast amount of work. Most attempts at television education have none of these. There is nothing worse than the results of a mediocre lecturer talking in a slightly embarrassed fashion into a television camera for an hour, especially if it is done in a highly expensive studio with blinding lights and cynical technicians who track the camera in and out.

The main value of education by television is that it can greatly enlarge the educational coverage that can be achieved by skilled teachers or unique authorities. Really good educators are few and far between. Technology can multiply their audience a thousandfold, or in the case of Alistair Cooke, a millionfold.

Perhaps the largest-scale use of educational television is Britain's Open University. Many thousands of students have obtained degrees from the Open University, attending lectures at home via radio and television. The students receive reading and activity packages by mail and send written homework by mail for marking by instructors or by computers. Counseling and tutoring services are available in person at several hundred local study centers. To impart a sense of belonging to an academic community, the students also attend one-week summer sessions.

The Open University has a campus, but unlike other universities it has no students on it. It is for the faculty. It houses the television studios and staff who create the lectures and pursue academic research.

Although it is very effective, the Open University is crippled by Britain's lack of broadcast television channel capacity. This gravely limits the number of lectures per night and restricts their viewing to certain hours, often causing family problems. The material developed in such universities could be used for decades around the world. Eventually, many full-time channels will be needed for such material. Even with the channel limitations the Open Universtiy grew in a very short time to become the largest university in Europe in terms of the numbers of students taking degree courses.

The use of television education *in industry* is growing rapidly, especially in North America. There is widespread use of video tapes and films for training, and many lecture rooms have closed-circuit television facilities so that lectures by authorities can be watched in other locations. In some cases long-distance video links are used for this purpose, but in most firms this is too expensive, because society has not yet built the video facilities it needs. The next generation of satellites can provide such links.

INTERACTIVE TELEVISION

There is a world of difference between lecturing to a camera and lecturing to an audience which can respond. This is especially true with a good lecturer who adapts sensitively to audience reaction. Lecturing via television, however, could be a two-way process in which the lecturer sees or hears the audience. As with most human communication, lecturing via a two-way electronic link can range from being abysmal to being a totally captivating experience. Much of its effectiveness depends on the design of the medium.

Some industrial and university links are designed so that a large audience watches on small monitor screens a lecture that is taking place elsewhere. The lecturer can see the remote audience, and the audience can ask questions. This is rarely very satisfactory, because the audience and lecturer cannot see each other *well*—the screens are too small or too distant. Members of the remote audience are usually reluctant to ask questions. Some organizations have used large screens as in a movie theater with projection television. This works much better, the large image having a more gripping impact on the audience.

The feeling of remoteness from the lecturer is removed if members of the audience are in small rooms with not more than eight people or if they are alone with a television set.

Imagine a class of thirty-two people, all in different locations and all using television sets. They may be in their homes or in business locations around the world. The teacher is in a studio with a bank of television screens on which he can see the faces of the audience. He can speak to any member of the class individually, and they can speak to him. They cannot see their fellow students and so tend to ask questions with little embarrassment from possible class reaction. The teacher may occasionally let the students see the rest of the class. If he wants, he may switch the face of a questioner onto the class screens. On the other hand, he can address any one student without others hearing. In addition to education, such a facility could be used for far-flung sales meetings or management briefing in a corporation.

Particularly important in remote education is the ability to use good graphics and charts, and for the instructor to be able to draw or write as he might in a classroom. IBM has a facility in New York in which the teacher sits in a specially designed television studio. The students sit at desks in small rooms, not more than eight students per room but up to ninety-six students in total. The students watch two color television screens, one that normally shows still images and one that can show the instructor, video tape, and more still images. The instructor cannot see the students but can hear their voices and observe collective reactions on a specially designed console. A student can "put his hand up" by pressing a button, and the teacher may ask him to speak. If the teacher leaves the student's microphone on he can interrupt verbally. The teacher can ask the students to react to statements or can give them multiple-choice questions. The students respond by pressing buttons. On his console the teacher sees a summary of the number of students responding and the number of responses in each category.

The teacher has complete control of three cameras. One is a movable camera with a zoom lens that has a very wide range of focal lengths. It can zoom in to the teacher's eyes or zoom out to show a picture of the whole room. Pointing vertically down at his U-shaped desk are two more cameras, also with zoom lenses, which can show charts, books, programs, or the teacher's handwritten notes or color drawings. The teacher also has control of a video tape machine. In front of him the teacher has four screens at about the same distance as they are from the student. He can see and adjust the output of the three cameras and the video tape. He can switch the contents of any of these four screens onto the students' two screens as he wishes. Also he can record the session and later analyze the effectiveness of his own teaching.

Teaching with this installation has proved to be very effective. The acid test is that the students, usually IBM customers, obtain better course grades than they would if they were in a conventional classroom. They find it easy to concentrate and tend to ask better questions than in a classroom. Some students in a conventional classroom tend to ask "showoff" questions subconsciously designed to display their knowledge to the class. They usually do not do this when the class is not physically present. The medium has a feeling of intimacy with the teacher. The cameras can close up on the teacher's eyes or on his fingers drawing or pointing to charts. The class can use a textbook and see the teacher turning its pages, pointing to words or diagrams and zooming in on details of interest. The students are usually seated in pairs and can confer with their neighbors without disturbing the class.

Not all teachers are effective with such a medium, but some are more effective than in a classroom. Some teachers have personalities that are best suited to a live classroom. Others prefer the controlled environment of the television studio in which they can point to documents, use photographs, talk quietly and intimately, see their own face on television while they lecture, and constantly monitor the class reaction statistically.

The equipment used in the IBM facility could be used effectively over a satellite channel or over cable television. If one still picture and one moving picture are transmitted, these could be sent together over *one* television channel. The still picture would be sent a slice at a time in the intervals between frames of the moving picture. The reverse channel requires the low capacity of a telephone channel and hence would be practicable on today's cable television systems. The system could be made to work with *broadcast* television in which the students have a telephone connection to the studio.

IBM experimented with a *video* reverse channel and found that it did not improve the effectiveness of the medium much. In fact, some students felt inhibited or uncomfortable because of the television camera pointing at them.

Society, with a little organization, could make massive use of this form of education. A small town with a cable television system could use it as a community educational facility. Universities employing satellite channels could make their best teachers available nationwide.

COMPUTER-ASSISTED INSTRUCTION

Computer-assisted instruction is an attempt to achieve *interactive* teaching in an automated fashion. Where it succeeds, the resulting program could be made available to *vast* numbers of people, like a successful movie.

As with video tape instruction or other canned media, the best computerized teaching is extremely effective, whereas the worst (today perhaps the majority) is insufferable. The difference lies in the skill and craftsmanship of the person producing the program.

Computer-assisted instruction has been used for a wide variety of people and a wide range of subject matter. It has been used at most levels of education, from very small children to graduate students and professionals. When the system is well designed, the pupils are captivated by the terminals; they learn at a fast rate with a high level of retention and finish each session with a sense of accomplishment. The computer is programmed to respond to them sensibly, with infinite patience, and with timing designed to maximize the reinforcement of the information in their minds. The pupils leave the terminals stimulated and often mentally fatigued. However, a badly designed system can be immensely tedious and can leave the pupils with their patience strained to the breaking point, thinking that it is a worthless gimmick and that they can learn much better from a book. To write well for this medium requires considerable care and talent. The writer must understand the subtle psychology of the medium.

Some topics lend themselves well to computerized teaching, but others do not. Subjects involving large amounts of routine details or facts are natural for computers, as are subjects with elaborate but standarized logical procedures. It is ideal for teaching spelling, simple mathematical techniques, the mechanics of foreign languages, statistics, computer programming, electronics, and so on. It would have more difficulty teaching philosophy, carpentry, basic principles of calculus, or appreciation of music, although even here it could assist a human teacher. In teaching a student to write good English, to draw, to play golf or a musical instrument, the computer is hardly worth considering. (Nevertheless, an IBM 1620 at Stanford University has been used to teach students to sing. This computer prints out a series of notes, and the student sings these to the beat of a metronome. The computer is programmed to compare the student's pitch with the true pitch and decide whether the student must repeat the exercise or go on to other material. *The Journal of Research in Musical Education* reported very favorable results, since the machine has a very precise sense of pitch.)

The computer is ideally suited for certain types of teaching but not for all teaching. It will never replace the human teacher. The teaching profession must recognize both the usefulness and the limitations of computers in education. No doubt as time and technology advance, the versatility of computerized teaching methods will increase.

Given a suitable subject and skillful programming, computer teaching can have significant advantages over conventional classroom teaching. Teaching in a classroom is almost always instructor-centered. The students have to proceed at the speed and level of complexity that is dictated by the instructor. The brighter students, finding the pace too slow, are bored; those who are not so bright often become lost or fail to understand part of what is said. Any student, bright or dull, can miss sections through lapses in attention. With computers, the process is pupil-centered, not instructor-centered, and the machine adapts its pace to that of the student. Dull students can ask for endless repetitions without embarrassment. Quick students or those who already partially know the material can skip a segment—with the machine questioning them to check that they do in fact know it. Students who want more practice in a certain area can obtain it.

With a well produced computer program, the student is unlikely to become bored or distracted. The machine will be programmed to elicit frequent responses and to grade the difficulty of its demands according to the ability displayed by the student. If a student is a long time replying to a question, the program will send messages to prod him or her. Students cannot race ahead without a true understanding of the material, because the program constantly "examines" them. What the machine lacks in human intuition, it makes up for in its tireless ability to provide a near-optimum response to almost every situation in the learning process. These responses have been carefully planned, with special regard to the psychology of learning, at the time a particular program is written. Furthermore, they can be easily revised as experience reveals deficiencies.

An elaborate program may have many branches and alternative paths. The responses to a question can elicit a variety of different actions. If correct, the main teaching routine will proceed. If partially correct or almost correct, a remedial sequence will be followed. If wrong, the instruction leading up to that question can be repeated or an alternative, more detailed sequence followed. If the answer reveals a lack of understanding of an earlier point, the program can backtrack. If the wrong answer is the culmination of many such errors, a branch can be made to a different teaching approach.

PROGRAM PRODUCTION

Writing these elaborately structured teaching programs requires much time and considerable skill. Furthermore, the program is likely to be modified substantially in response to the reactions of its users. For this reason, the computer often records the users' answers. A printout shows the author the students' exact replies to each question and indicates how long they delayed in answering. The author may then make many modifications to the program. In addition to this feedback, the author normally wants to talk to the students and observe their reactions to the terminal. Unlike a movie or video tape, the computer routine can undergo endless modification and polishing not only by the author but also by the teachers who use it. Furthermore, when a program acquires an outstanding reputation, as a few have, it can be duplicated, studied, and used in machines throughout the world.

An outstanding program for computer-assisted instruction is a work of great art. A strong sense of style is needed. This is a new medium quite different from any that have preceded it, and needs its own rules concerning style. As yet, no acknowledged sense of style has developed. The medium is too new. One can read books on literary style, but not on this. No doubt a style guide will be developed in time, and it will probably change with time just as style in other media, such as movies, has changed. Computer-assisted instruction may be much more susceptible to change because of the intimate two-way interaction, which no other medium has, and because of the immense programming versatility that is possible.

In the meantime some singularly unstylish programs are being written. Programmers, hurriedly attempting to demonstrate the new machines and infatuated by the ease with which they can make their words appear on the screen, are producing programs as bad as the home movies of an amateur with his first 8-millimeter, zoom-lens, camera. Such programs often lack the thoroughness which the medium demands in reacting to student responses. This is probably a temporary dilemma; teachers rather than programmers are taking over the programming work as the technology spreads.

It would be unfortunate if the public or the teaching profession (most of which is avidly looking for reasons to hate computers) formed their opinions on the basis of the bad programs that are now circulating. A reasonable judgment of the potential of the medium should rest on the handful of *exceedingly* effective teaching programs that have been created; but these are submerged in the welter of shoddy products of average programmers. Eventually, the production of teaching programs will become as elaborate and professional as current motion picture production; today's best programs will seem as crude then as pre-World War I movies seem now.

IN THE HOME

There will be a massive growth in computerized teaching when it becomes widely available in the home. This is beginning to happen, not via telecommunications but by use of small electronic units that are plugged into the antenna leads of conventional television sets, like the television games that are now being marketed. These units—microminiature computers—use small, interchangeable magnetic tape cassettes like those used on a cassette recorder. The cassettes, or other storage media, contain data which the microcomputer can select and display on the television screens.

Telecommunications links may later be used with the domestic television set, to gain access to what will eventually become huge libraries of computer-assisted instruction programs and data banks. Such facilities could be provided via cable television links. In the future a community with good schools may be expected to provide good computerized education via the community cables. Children will learn and do their homework with this facility, under direction of the local schools.

THE EDUCATION INDUSTRY

There are three levels at which computer-assisted instruction can operate, differing in channel requirements.

The lowest level displays alphabetic text on the terminal. This, like all uses of computer dialogues with text, letters, and numbers, has low bit transmission requirements compared with digitized telephone transmission.

The medium can be more effective and stimulating when pictures are used. The second level can display still color pictures on the screen. Some systems have used two screens, one for text and one for pictures. Substantially more bits are required to store or transmit pictures, but the requirement is low compared with the transmission capability of cable television.

The third level uses movies. An educational movie will stop at intervals to show still frames and ask the student questions. The response to the questions will determine what frames are shown next and whether the machine repeats or skips segments of movie. The selection of the next movie segment may take ten or twenty seconds, and this selection delay is hidden by showing still frames. Here we are talking about more expensive equipment, possibly using a video machine instead of a data cassette, or possibly using a television channel from a central computerized facility. When this technique is used well, the results can be gripping and powerful.

It cannot be stressed too strongly that the production of good computer-assisted instruction is a highly skilled, professional operation. One day a substantial industry will grow up to produce such programs. It may become an industry the size of Hollywood and just as professional. More than one thousand man-hours of work will go into preparing one hour of top-quality instruction.

A glimpse of this can now be seen with non-interactive education in the Open University campus and the studios of video education companies. The best that is produced will exist for many decades, as do classic books, recordings, and films. It will be part of the cultural heritage of the satellite age.

Today when so many difficult problems face humankind the development of powerful and wide-ranging thinking is desperately needed. In the past, Britain has produced some remarkable scholars gifted over a wide range of disciplines. Such individuals are sometimes referred to as polymaths. In recent times they include Bertrand Russell, A. N. Whitehead, Arthur Eddington, J. B. S. Haldane, Arnold Toynbee, and Jacob Bronowski. Bertrand Russell commented that a childhood period with a wide diversity of interests is needed for the development of such individuals, with no pressure for conformity. The child should develop and pursue its own interests for a period no matter how unusual they may be.

Countries or educational systems with strong pressures for conformity will produce technicians but few polymaths. Television has an immense influence on children. It could be a force for conformity, a trivial medium for idle distraction, or a medium which encourages learning across an immensely broad and diverse range. Which of these it is depends upon the technology, how much money is spent on the cables, and what laws encourage or restrict the development of the medium.

DEVELOPING NATIONS

Educational television will be exceptionally important in developing nations, where lack of skilled manpower is one of the major factors inhibiting growth and there is an inadequate supply of teachers. The faster a nation is developing, the more serious are these shortages. One of the fastest developing nations is Iran. Iran expects to develop its industry so that its gross national product rises from $570 per person in 1972 to nearly $5000 by 1995. Because Iran's oil revenues will have fallen to less than $500 per person by 1995, Iran must undergo massive industrial development over a twenty-year period in order to meet its goal. These projections make a major assumption: that sufficient skilled manpower will be available. It would be impossible to train the manpower required with conventional forms of education and training.

The mix of manpower in developing countries is entirely different from that in developed countries. This can be seen by comparing Iran in 1966 with France, Germany, England, or the United States in Table 19.1.[2]

Table 19.1

Percentage of the working force in the following categories:	*France (1968)*	*Germany (1970)*	*United Kingdom (1971)*	*United States (1973)*	*Iran (1966)*	*Iran (1972)*	*Iran estimates for 1992*
Professional, technical and related	11.4	9.8	11.1	13.2	2.7	4.6	11
Administrative and managerial	2.7	2.2	3.8	9.6	0.1	0.2	4
Clerical and related	11.7	17.6	17.9	16.7	2.7	2.6	16
Sales	7.6	8.9	9.0	6.2	6.7	7.8	9
Service	8.4	9.5	11.8	13.0	6.7	11.6	12
Agriculture, animal husbandry, forestry, fishing hunting	15.3	7.7	3.0	3.4	41.3	39.6	11
Production, transportation, laborers	34.6	36.3	39.9	34.6	26.8	33.3	37

Iran is rapidly training professional, technical, administrative, and managerial personnel, but a more rapid rate of training is needed. Table 19.1 shows how the proportions of the workforce must change. A country of 40 million people needs to train more than one million professional and technical personnel, two million clerical personnel, and two thirds of a million managers and administrators, in a period of twenty years. At the beginning of this twenty-year period 86 percent of the rural population was illiterate. In addition to school training, adult literacy training is needed at a rate of 1.5 million persons per year. To achieve long-term development over, say, forty years, it is desirable to improve the education of children. To achieve rapid development over, say, fifteen years as desired in Iran, it is necessary to tackle the training of adults. In Iran this must occur on a massive scale. The schools, colleges, and teacher resources are quite inadequate for the volume of training needed and must be supplemented by electronics.

There have been many uses of educational television in developed countries. This experience has shown that its use with children brings only a small change in the desirable student–teacher ratio. Its advantages (when designed with quality) can be enormous enrichment of curriculum, improvement of student interest, and more effective homework assignments. The same is true with computer-assisted instruction: The teachers are still needed.

In adult or post-secondary education, however, computer-assisted instruction can be used effectively with a much smaller number of teachers, and with no teachers if the students are self-motivated and the program quality is good. At Oregon State University one instructor handled an enrollment of 700 in a biology class, using television and twenty-two teaching assistants who were advanced undergraduates or graduate students spending eight hours per week on the course, and each handling groups of thirty to thirty-five students. The results were good. This pattern seems appropriate to the universities of developing nations, and some have used it with great success. The University of Pahlavi in Iran conducted three courses using educational television; 2000 students completed the courses, and 1800 passed the exams.[3]

Many areas of developing nations have been exposed neither to intensive education nor to television. The social and religious leaders are often opposed to the introduction of new ideas, and organizations from outside the local community are rarely accepted. To succeed, education must be conducted by members of the local community. Assistant teachers—mentors—selected from the local community are being trained to work with the television in Iran, to relate what it says to local problems, and to overcome local resistance as far as possible. In Iran, even doctors have found extreme difficulty being accepted in rural communities. The choice of respected community members as mentors is critical to the success of educational schemes.

In rural areas of Iran, like most developing nations, the postal services are poor or nonexistent. To educate the nation with television, many thousands of hours of television will have to be distributed to many thousands of communities. The least expensive way of doing this is via satellite.

A satellite system has been designed for Iran to provide television distribution along with telephone, radio, data, and facsimile transmission. It will have large earth stations which can *transmit* television in sixteen major cities, and small earth stations which can only receive it in 12,000 smaller communities that constitute two thirds to three quarters of the population. The telephone, data, and facsimile channels are important for enabling the local mentors to interact with the principal instructors, and handle administrative functions. The data channels can be used for computer-assisted instruction.

Some television may be used live from the satellite, but much will be locally recorded so that it can be used at times suitable to local operations, and possibly so that the sound track can be changed to local dialects or languages. Much printed matter will be distributed to back up the video programs.

The task Iran has set itself, of developing industry comparable to Europe's by the time its oil revenues decline, is awesome. It will be interesting to observe how effective telecommunications-assisted education proves to be, and how soon similar techniques are used by other developing nations.

DEVELOPING CORPORATIONS

Just as some nations are developing at a rapid rate, so are some corporations. Education of a large number of staff will have a major effect on their future profitability. Both computer-assisted instruction and television education are being used effectively in some corporations; however, television via telecommunications has little or no use yet, because the cost of video channels is too high. Satellites will eventually bring the cost down to an acceptable level. The same links that are used for teleconferencing may be used for video education, perhaps using satellites such as those to be launched by Satellite Business Systems.[4] Video education could be distributed much less expensively with one-way television channels like those to the rural communities of Iran. Many corporations would make use of such a common-carrier or national facility.

As with developing communities, the local corporate offices would have an individual responsible for education to act as a local mentor and ensure that the facilities were used correctly.

Chapter Twenty

1984

The maximize the chance of freedom tomorrow, we should build the greatest diversity *of information channels today.*

In 1949 George Orwell produced a vision of society with telecommunications that has haunted the electronics industry ever since. The vision might have been worse if he had known about computers:

> Behind Winston's back the voice from the telescreen was still babbling away about pig iron and the overfulfilment of the Ninth Three-Year Plan. The telescreen received and transmitted simultaneously. Any sound that Winston made, above the level of a very low whisper, would be picked up by it; moreover, so long as he remained within the field of vision which the metal plaque commanded, he could be seen as well as heard. There was of course no way of knowing whether you were being watched at any given moment. How often, or on what system, the Thought Police plugged in on any individual wire was guesswork. It was even conceivable that they watched everybody all the time. But at any rate they could plug in your wire whenever they wanted to. You had to live—did live, from habit that became instinct—in the assumption that every sound you made was overheard, and, except in darkness, every movement scrutinized.
>
> Winston kept his back turned to the telescreen. It was safer; though, as he well knew, even a back can be revealing. A kilometer away the Ministry of Truth, his place of work, towered vast and white above the grimy landscape. This, he thought with a sort of vague distaste—this was London. He tried to squeeze out some childhood memory that should tell him whether London had always been quite like this.[1]

Orwell's telescreen is still a little beyond the capabilities of the British Post Office, but there are causes for concern well within the scope of today's technology. First, bugging devices are now so small and efficient that electronic eavesdropping on speech is easy. Second, many different computers collect data about an individual. In total the data they store could form a formidible and highly personal dossier if it were gathered together. Third, the power of modern television is immense, and an authoritarian regime could use it as a tool of persuasion, possibly eliminating alternative information channels. An authoritarian regime could use the media to impose its views on the public and use bugs and computers to carry out surveillance of the public to detect dissenters.

CONTROL OF THE NEWS

Many politicians in the past have realized that a major step toward increasing their power would be control of the news. Many have commented that this step is more potent with modern electronics, because television makes the news more gripping and emotional and gives vast, immediate, nation-wide coverage. The more powerful the medium, the more effective it can be to a politician who gains control of it.

The speed and power of television and radio is awe-some. When President Kennedy was shot, 44 percent of Americans knew about it within fifteen minutes, 90 percent within one hour. The percentages were only slightly lower in many foreign countries. The events following the assassina-tion were seen by 96 percent of all American homes and were watched, on the average, for 31 hours and 38 minutes. When President Nixon announced his decision to have a truce in Vietnam in 1973, he was watched by 62 percent of all U.S. households. The first Ford–Carter debate before the 1976 Presidential election was watched by 73 percent. The launch of Apollo 17, the last U.S. moon shot, was watched by 81 percent of all U.S. households.[2] If Europe had had such television, the dramatic events prior to World War II would have commanded similar audiences. Would this have changed the course of history? With audience response tele-vision, would the same catastrophic decisions have been made?

Politicians using television are face to face with each viewer, employing a script and image meticulously planned by experts. No individual in history, not Alexander the Great, Julius Caesar, Napoleon, nor Stalin, had this com-munication power.

In America perhaps more than any other nation, enormous talent has been spent on the art of persuasion via television. It has not been used on social and political issues, but on selling commercial products. More than $7 billion per year are spent on television advertising. If an authoritarian regime used similar talent to spread its social and political views, the results would be devastating. Historians of the future may regard it as an amazing phenomenon that with American television sets switched on 6½ hours per day there is so little political and social use of the medium. *Ten* corporations that manufacture mass-consumption goods pay for 78 percent of all advertising.[3] and hence control almost that proportion of what average Americans view. These and other corporations control more input into American minds than all the schools, churches, and politicians put together. This use of media has created humankind's busiest, richest, supermarket culture. The same power could be used by others to create a different culture—better or worse.

DIVERSITY

If communications technology is further expanded as described in this book, will that increase the danger of an authoritarian regime imposing its views on the public?

In fact, authoritarian governments abhor *too much* telecommunications. They appreciate the power of modern media but want to limit communication to *controllable* channels. Authoritarian governments want a small set of channels from a central authority to the people, and little or no lateral communication.

In the early 1970s there were fewer telephones per person in the U.S.S.R. than in Fiji. Less mail is sent in the U.S.S.R. than in New York City. Hitler banned telephone conference calls because they might make possible unseen and undesirable meetings.

Not surprisingly, the Soviet Union was a pioneer in television, with its first telecasts made as early as 1931. Today there are few channels and no *local* television stations. Most totalitarian countries have only one television channel. The Soviet Union leads the world in the technology of newspaper production and distribution. It has 2½ times the geographic area of the United States, yet most of its newspapers originate in Moscow (which has 3 percent of the population) and are produced and distributed with greater speed than in the West. Local newspapers and local television stations that can determine their own programs are largely avoided in totalitarian countries.

Dr. Klaus Vieweg, an authority on mass communication from East Germany, describes the communications media of the communist countries as follows.[4]

> There is no private ownership of printing presses, electronic media equipment or other similar institutions. The material base for mass communication is owned by the people; the government, political parties or mass organizations such as the trade unions decide about the content of communication. These and other principles contribute to the mental enrichment and self-education of the audience and to a better and deeper knowledge of the world.
>
> In capitalist society, the functions of journalism are different. There are consequently different conceptions concerning theoretical and practical questions of journalism and mass communication, conceptions which lead to almost complete divergence of ideas regarding, e.g., news value.

In 1950, 88 percent of all radio receivers in the Soviet Union were *non-tuneable.* Many were wired systems in which loudspeakers in apartments were connected to master receivers. The listener could obtain only one station and in some cases could not switch it off. More recently the number of tunable, "free" receivers has increased.

In Eastern European socialist countries, one third to one half of all television output is devoted to news, current affairs, documentaries, and education. Television authorities refer to "nation-building" programs. Feature films, series, and entertainment take one third or less of the program time. In some of these countries much emphasis is put on children's programs. In China 90 percent of programs are news, current affairs, documentaries, and education, and 10 percent are children's programs.

An authoritarian government may want the power of

modern media to impose its views on the public or to spread deceptive information; it may also want to restrict the channels to prevent some types of knowledge from being passed on. One of the government slogans in Orwell's *1984* was IGNORANCE IS STRENGTH.

Free flow of information can be most hazardous for an authoritarian regime. Even nontotalitarian governments can be concerned with the free flow of some types of information. A leading member of the British Labor Party commented in 1974 that the free flow of information in society is as undesirable as the free flow of sewage.

The danger of restricted information flow is that the government must be blindly trusted to make those decisions which are best for the society. The evidence of history is that governments are unlikely to succeed in this. The public have rarely benefited from prolonged dictatorships or authoritarian control.

The communications technology in a society has a major effect on how that society can be governed. If the channels are primarily *from* the government *to* the people, then control of information is practical. The greater the quantity and diversity of *decentralized* and *lateral* channels, the greater the potential number of information sources, and the more difficult it becomes to limit and control the flow of information. Modern trends in telecommunications have tended to make the insulation from free information increasingly porous.

To ensure a free flow of information, there appears to be a principle relating to the design of a society's communications technology. *A free society should have the maximum diversity of communications systems.* The greater the diversity, the more difficult the restriction and permanent falsification of information.

A free society should have local as well as national television and radio channels. There should be many sources of programs originating in different types of organizations, and it should be possible to distribute these programs freely across the society. There should be cable television as well as broadcast television and direct broadcasting from satellites. There should be newspapers and magazines as well as television and radio. For this reason the reduction in numbers of newspapers is an unfortunate trend. *The New York Times* needs competition. We should not talk, as some McLuhanites do, about video media *replacing* print. There should be good national distribution of printed news as well as local news sources. Computer technology should provide more detailed, more accessible, and well indexed information sources. We can make these available from the home, and this should be done. The "gatekeepers" should package the news but not throw it away as they do today. Reporters should not be restricted by the teletype channels or other bottlenecks in what they write. There should be broadcasting of data as well as of sound and television. Television should have both commercial and noncommercial channels. It should carry both glossy, highly edited news and longer, detailed news programs. There should be direct broadcasting of government meetings and debates. There should be many television channels, and they should differ in nature. The imperative to maximize the audience should apply to only a few of them. Consistency of style between channels is undesirable; there should be as much diversity of video information as there is of print.

Lateral communication channels between people are important. The telephone is the main lateral channel in advanced societies today. It should be made easy to have conference calls, to have speaker-phones which permit a room full of people to hear and talk, and to have telephone message-leaving facilities. Satellites and other technology should be permitted to bring down the cost of long-distance telephone calls. Citizens' Band (CB) radio adds a new and important type of channel. Its technology should be developed, and it should be made possible to dial telephone calls via CB radio. Electronic mail and message networks should be built. The data networks should be accessible from the home both to obtain information and to send messages. Data network conferences, with participants at home, could be an important type of lateral communication. Lateral video channels should be built when the cost of the technology falls sufficiently.

News and current affairs programs from other countries should be made available. These should be given several sound channels in different languages. They may be distributed via satellite.

Max Lerner, on completing his massive two-volume work *America as a Civilization*,[5] was asked by a group of foreign professors if he could select any one word that epitomized what was special about America as a civilization. After some thought the word he selected was "access." It is a country in which people have *access* to information, to management, to education, to opportunity, to politicians. It is perhaps pertinent that America has generally had the world's best telecommunications. Totalitarian countries have generally had the worst.

To maximize the chance of freedom tomorrow, we should build the greatest diversity of information channels today.

Chapter Twenty-One

THE HUMAN GOLDFISH

A society with a high level of automation must eventually frame its laws and safeguards in such a way that computers can police the actions of other computers.

We commented earlier that polar bears could become extinct. Some polar bears have another minor problem related to this one. They have embedded in their flesh a miniature radio transmitter that transmit periodic signals to an earth satellite. The signals reach a computer, which analyzes the movements of polar bears and summarizes them for researchers.

One common concern about a telecommunications society is that computers could keep track of people just as easily. Today many different machines contain many different pieces of information about us. Electronic bugs are tiny, powerful, and for sale at low prices on Broadway. Electronics could make our lives as visible to officials as that of a goldfish in a bowl.

The trouble is, democracy flourishes on information. That information should be good; it should be accurate, and it should be balanced. For that matter, a nondemocratic society needs information too, though often of a different sort—the government may need information in order to protect itself. A democratic society should be sensitive to the needs and welfare of its citizens, and to look after these, it requires information about its citizens. The more complex the society and its technology, the greater the need for information. Herein lies the dilemma.

The problem with "privacy" is its conflict with other social values, such as competent government, a free press, protection against crime, health care, provision of services, collection of taxes, social and medical research, and the development of community living environments. The authority providing each of these wants to decide what it should know about us and when it should be told. We, on the other hand, resent the intruding official eye.

To date, democracy has had to make do with rather poor information—distorted, partial, often nonexistent. There has been a tradeoff between speed and accuracy. However, we have muddled along in a tolerable manner. In places the ill effects of poor information and insufficient analysis are evident, sometimes catastrophically so. Many of America's appalling urban problems could have been lessened, given good information and government that could take the right corrective action.

Today we have at our disposal new machinery of great power. We can collect, transmit, and analyze thousands of times the information we ever could before. The conflict between privacy and other social values suddenly becomes acute.

PERSONAL INFORMATION

There are currently in operation many varieties of computer data banks holding personal information. They include the following:

Police
- FBI
- Security clearances
- Police information systems

Regulatory
- Tax
- Licensing
- Vehicles

Planning
- Property owners
- Vehicles
- Economic data
- Business information

Welfare
- Medical
- Educational
- Veterans
- Job openings and unemployment

Financial
- Credit bureaus
- Savings and loan associations
- Banks

Market
- Mailing lists

Organizational
- Personnel files
- Membership lists

Professional bodies
Armed forces
Corporation employee dossiers recording intelligence, aptitude, and personality tests, and appraisals and attitudes

Social
Computerized dating
Marriage bureaus
Hobbyist data

Research
Medical case histories
Drug usage
Psychiatric and mental health records

Travel
Airline reservations with full passenger details
Hotel reservations
Car rentals

Service
Libraries
Information-retrieval profiles
Insurance records

Qualifications
Education records
Professional expertise
Membership of professional groups
Results of IQ and aptitude tests

In each case sound, cogent reasons have been evinced for collecting and retaining the data in these banks. Few of these data banks, however, guarantee confidentiality or anonymity.

There are those who advocate that the public's *right to know* about an individual should take precedence over the individual's *right to be private.* An amazing number of government officials claim this about one aspect or another of the individual's life. Some see belief in the right to privacy as an elitist attitude, especially where it involves information about rich or powerful people.

Information that could destroy one person's career might be a matter of indifference to another. In England at the turn of the century, practical joking was in fashion. A notorious prankster sent telegrams to all the bishops and deacons of the Church of England, saying "Fly at once. All is discovered." Seventeen of them caught the next boat to France.

CONSENT

Some computer systems collect information with the *consent*, expressed or implied, of the individual in question. In other cases the public would be highly alarmed if they knew the details that were stored about them. The credit bureau computers, for example, do not merely collect *financial* information. Some collect data under the heading of "moral hazards," including extramarital affairs, homosexuality, heavy drinking, and other social observations which are made that might affect a person's credit risk. Some of the information is collected by the most dubious methods. A vicious comment from an apartment doorman who was not tipped could lead to a black mark on the record. Dossiers are kept on more than half of all Americans.

While many Americans expect to be on the credit bureau files, few would expect to be on the Department of Defense files of potentially subversive or harmful citizens. However, a Senate subcommittee chaired by Senator Sam Ervin in 1971 revealed that the Central Index of Investigations in the Defense Department has 25 million cards, each relating to a person (one sixth of all adult Americans and a very high proportion of Americans in the age range eighteen to twenty-six), and 760,000 relating to organizations and incidents in the United States. On an average day, 12,000 requests for information in this file are processed and 20,000 additions and changes were made. Former agents testified to the committee that information they collected was often passed on to other government departments and local police. Constant surveillance was maintained of the American Civil Liberties Union and other bodies such as the National Association for the Advancement of Colored People. A court ruled that this data bank should be destroyed; however, two years after it was said to have been destroyed, a small news story described its being transmitted to a different location on a data network.

England has a scheme to keep continually updated computer records of the entire land surface of Great Briatain on a grid system with squares of *one meter*. One wonders how such fine-scale information could eventually be used.

A wide variety of information about individuals is sold for marketing purposes. A New Jersey firm has a project to market to drug companies computerized data on the buying habits and personal backgrounds of doctors. Tapes of data listing names and addresses of persons with special interests can be purchased. A Manhattan church requested its attendees to fill in a card if they are interested in its Bach concerts; persons who did so were bombarded with details about other concerts and records, the church presumably having sold the information. The computerized dating firms have sold tapes giving details of what their subscribers filled in on the forms they were sent.

Several city and local governments in the United States have plans to establish computer-stored dossiers on every citizen in their areas. More ambitious was a proposal for a National Data Center, combining the files of many separate authorities. A furor in the press and Congress resulted in the National Data Center plans being shelved, at least for the time being. The local data banks are going ahead, and they are less well protected than a National Data Center could have been.

The information which government departments collect can easily be transmitted to other departments. If data processing operations were made as efficient as possible, there would be much sharing of data. The chairman of the U.S. Civil Service Commission said,

> For proper decisions we must have integrated information systems. This will require the use of information across departmental boundaries. It is here that current efforts to standardize symbols and codes will pay dividends. Direct tape-to-tape feeding of data from one department to another may become common. These systems will mesh well with developing plans for an

> executive-level staffing program which will be designed to locate the best possible man for any given top-level assignment, no matter where in government he may be serving.
>
> The computer's ability to search its perfect memory and pick out records of individuals with specific characteristics has been applied in the search for candidates for Presidential appointments. A computerized file containing the names and employment data of some 25,000 persons, all considered likely prospects for federal appointive positions, is searched electronically. The talent bank, with its automated retrieval system, broadens the field of consideration for the President in critical decisions of leadership selection.

The dossiers kept about individuals often contain information that even the subject himself does not know. The New York State Identification and Intelligence System keeps detailed criminal, social, and *modus operandi* files on criminals and *suspects.*

There is justification for holding extensive data on known criminals, or suspected criminals, or suspected spies, or potential criminals, or potential spies, or people who associate with known criminals or work on secret projects for the government, or for a government subcontractor, or . . . It is difficult to draw a line.

INEFFICIENCY

In the past, a primary safeguard on privacy has been inefficiency. Government records have been uncollated, uncentralized, erratic, inaccessible, difficult to interpret, and often erroneous. It was expensive to collect and store data. The separate files were not interconnected or cross-referenced. With computers, that is changing. The vast electronic files, automatic indexing, and telecommunications links provide a tool with which any amount of data can be collected, stored, and made available. There are economic arguments for interlinking financial, medical, insurance, license, welfare, marketing, and organizational records.

SURVEILLANCE

During the Vietnam war, South Vietnam was plagued by black-market currency operations. To check them, computers were programmed to detect unusual activity or balances in the bank accounts of Americans.

The computer is a relentless, unforgetting instrument for keeping watch on such activities. What can be done in Vietnam can be done at home, and it can sometimes be argued that this would be good for society. Surveillance to spot tax evasion, on the face of it, sounds like a good idea. Programs in computerized telephone exchanges for tracing malicious anonymous callers (a system in use in Sweden) are highly commendable. Computerized surveillance to help crime detection appears to be potentially valuable. But how far should this be taken? Most citizens would react indignantly to the proposal that a computer monitor their bank account or credit transactions for unusual activity. In fact, checks over a certain value are automatically recorded in the United States. How would we feel if we knew a computer were analyzing all the telephone numbers we dial as an aid to national security or crime detection or for some obscure bureaucratic motive?

Electronic surveillance could be used to keep watch on telegrams, bank account postings, credit transactions, licenses, stock market purchases, airline travel, hotel bookings, car rentals, and so on. Invariably, some official somewhere has a good reason for using such information. The police want to know when such-and-such, a suspect, makes an airline booking, or if so-and-so attempts to enter the country or to check in at a hotel in Buffalo. Certain Social Security numbers belong to people who have evaded bill payment; watch out for their next appearance. What bank accounts in the Phoenix area showed unusual activity during the week beginning May 13? The more information about our daily lives that is stored in computers, the more easily such questions can be answered automatically.

Surveillance is a logical extension of automating the tax system. In the future the IRS may well monitor even more of our financial life; it might monitor some individuals totally.

Somewhere in different machines an enormous amount of information about us will be available—about our financial, educational, medical, business, and legal affairs, our aptitudes, magazine subscriptions, library use, and so on. Most of it will be collected without anybody obviously prying into our affairs, as a byproduct of other mechanisms in society. The extent to which the machines or agencies can intercommunicate to bring the facts together will determine how complete a picture of an individual's existence can be constructed and analyzed.

The use of credit and the growth of electronic fund transfer, described in Chapter 9, will leave a spectacular trail of computerized information. In a society with a great reduction in the use of cash, many of the things we do will be recorded in the form of computer transactions. Paying the thruway toll as we leave home, buying gasoline and lunch, paying for a hotel room and drinks in the evening, buying presents—all of these will be recorded because we use our credit cards. And someone will have reasons for analyzing it, if he is allowed to. People who want to sell us things want to know of our spending habits and will pay for listings of consumers in different categories. Conducting such surveillance has been out of the question in the past because of the overwhelming amount of work involved. With computers it could be made easy.

All manner of organizations can concoct reasons for wanting the results of surveillance. Marketing executives want to spot potential customers. Police want to spot potential criminals. Insurance companies want to detect bad risks. Cities want to spot habitual parking offenders. Politicians want to locate potential supporters. Corporations want to spot potential managers. Shops and credit companies want to spot credit risks. Private detectives want information about their suspects.

BREAK-INS

The prospects of *authorized* utilization of the contents of the data banks is alarming enough. A further cause for alarm is that in some cases individuals have gained *unauthorized* access to the files.

Perhaps some future Perry Mason dialogue will read something like this:[2]

> "Okay, Perry. I've identified the guy's girlfriend—the one who's getting his insurance money."
>
> "How in the world did you get a line on him, Paul?"
>
> "Through the terminal in the local office of Poli-State Life. I asked the little blonde to get me a quote for a policy and as she keyed in the data, one of my men, Harry, was around back listening on the wires. When I saw the rates, I accepted and paid her $100—you'll see it on your account soon enough—and she filed it away. Then I said: 'Did it cover private aviation?' And she said, 'No that's extra.' So I got her to quote again and correct the cover—another $28—and Harry recorded her dragging the record out of the data bank to look it up."
>
> "Say, wait a minute, Paul. I see how you could get your record out, but how did you get Winston Smith's?"
>
> "The operator who was there took his recorder home and decoded everything—none of it was in cipher and all the characters were in USASCII. . . ."
>
> "Is that the standard code for information interchange I was reading about?"
>
> "Uh-huh. It was a dial-up line and the program is very simple. All the girl at the terminal does to look at an item of data in a record is key in the number of the field in the record. So we dialed up the Poli-State computer from a rigged terminal with the right response circuitry, pushed in his policy number, . . ."
>
> "Where did you get that?"
>
> "From his bank statement—it's direct-debited. We just kept to the drill the blonde used for my policy, and then tried every number from 1 to 32 when we struck oil. Beneficiary's name and address was typed out. So we said thank you, goodbye, and the computer wrote the time and signed off. Marvelous things, these computers."

Paul Drake and Perry Mason are known for their ingenuity—or at least for that of Erle Stanley Gardner—and their ability to maneuver within the shoals of the law. Very few computer users are in a position to outwit them and their nonfictional ilk. Even fewer have bothered to try. Data banks can be made robber-proof—at a cost. The more ingenious the potential robber, the greater the cost. The technology for data bank security is now understood.[3] However, most of today's data banks are grossly insecure—most of them are not protected at all.

It seems certain that in the future we are going to live out our lives against a background of highly informed and intercommunicating computers. For the decades ahead the data banks will continue to grow and multiply. Financial, police, and government networks are only some of the innumerable networks that will come into existence. Computer terminals will be everywhere, and gaining access to the appropriate machine for a particular function will be almost as easy as dialing on the telephone today.

Many of the systems will be set up with highly altruistic motives, like the systems that saved the car crash victim at the start of Chapter 3. Government data banks will be intended for better planning. Medical data banks will be intended for better research to attack such problems as cancer and mental illness. Police data banks will be intended for better crime control. However, the data banks that are so beneficial in one circumstance can constitute an invasion of privacy in another. The electronic society has to steer a delicate course: on the one hand making the best use of the machines, and on the other hand limiting their adverse effects.

If automation is used well it can have many benefits. We will probably have more leisure time, fewer routine chores, perhaps fewer forms to fill in, greater affluence, more intellectual pastimes, perhaps less chance of being bored, and more opportunity for creativity. A person with the intellectual capabilities demanded by the new age will have a greater chance of achieving something worthwhile. A person without them will probably seek activities unrelated to those which require logic. But we will always be aware of the machines. They will be part of society—their knowledge of us, of our telephone bills and medical histories, our educational attainments and consumer habits, and so on. More of our mail will come from computers than from people—computers trying to sell us things, sending us bills, asking for money for charities, notifying us of bank debits for their services, and so on.

To what extent could this constant electronic activity constitute harassment? To what extent will telecommunications intrude on our desire to be left alone? This depends on the laws, regulations, and social attitudes relating to privacy.

PROTECTION

If societies of the future are to be decent civilized environments, citizens need protection from electronic invasion of privacy. This is true whether or not advanced telecommunications links are built, because all the computer applications mentioned in this chapter are being implemented anyway with today's telephone lines.

To ensure privacy three steps must be taken. First, the computers and data banks must be designed to be secure. Second, there must be laws that limit and control the use of systems and bugs, and those laws need to be enforced so as to ensure privacy. Third, the societal attitude to privacy must be such that there is a major outcry when violations occur, telephones are tapped, bugging devices are used, and so on.

The *machines* can be designed so that they are secure, reliable, and accurate and so that they protect private data. It is important to realize that data in a computer *can* be protected securely—at a cost. Some press articles have implied that any person at a terminal can browse indiscriminately in the computer files. This is not so. On a well designed system he can gain access only to the data intended for him. Bank vaults are locked securely, and military compounds have multiple layers of protection; the same can be true with computers. The means of protection are not always brute-force measures like guards and strong rooms but instead can be electronic locks and alarms. Unlike their human counterparts, electronic securities can be relentlessly thorough. Data can be further protected if so desired by cryptography, as with the messages transmitted by spies, and because a computer does the enciphering the codes will not be broken easily. Such safeguards are not highly expensive, but their cost will vary with the degree of security needed. Protection against ordinary snoopers is easily achieved. Protection against James Bond or the CIA needs more ingenious and expensive measures. A fairly elaborate set of safeguards might add, say, 5 percent to the cost of the system.

Today many computer systems do not apply these protective measures. In the case of some computer installations, this is because of lack of forethought. (It is staggering to consider that the early police information systems were designed in such a way that an intruder with a terminal one thousand miles away could—and sometimes did—break into the files.) In other systems. it is because the owner of the system saw insufficient returns from the expenditure of 5 percent of the system cost on protecting clients. Occasionally the owner had much to gain from selling private information about clients. In such cases, both an appropriate legal structure and the capability to build electronic locks are essential.

Electronic protection of data in computers will eventually become a standard part of the technology. It will be easier, cheaper, and more efficient than today. Computer personnel should be urged to pay attention to such safeguards; otherwise, they may find that public and congressional outcry will prevent them from building their systems.

NEW LAWS

There have been many proposals for legislation.[4] Many bills have been drafted in the United States and other countries. The proposals range from constitutional amendments to administrative regulations. Although it is easier to say what is required than to determine how to enact relevant legislation, the general form of some of the laws needed is clear.

Corporations are obliged to adhere to certain regulations governing their mode of business. Accounts must be rendered in terms set down by law; workshops and offices must meet health and safety standards; contracts are established within a known framework. Such a legal system has grown in answer to needs, sometimes as a result of blatant injustice. We must establish a legal environment for information systems now to avoid injustices we can already predict. Good systems design, like proper accounting, can be legally enforced.

In the United States the bills that have been drafted or passed tend to deal with isolated aspects of the problem, such as use of information by government agencies, fair credit reporting, and the collection of data on citizens by the Department of Defense. In some other countries more general bills have been drafted. Some bills seek to establish a register of all data banks to control their use.

A number of general comments can be made about the legislative proposals. First, there must be recognition that modern technology allows one person to gain without another necessarily losing. Second, an extension of the concept of private property does not provide a suitable framework for data privacy laws. The computers copy information, so that nothing physical need be taken.

Third, it seems as important to have legislation concerning *private* data banks as about public or government data banks. Many horror stories arise from privately owned data banks such as those of credit bureaus, employment agencies, and private detection agencies. It seems reasonable that, like automobiles, computers should have to meet certain basic safety regulations. Guidlines for corporate regulations on the use of computerized data ought to be maintained by the professional bodies. However, inadequate safegards that result in disclosure of sensitive personal data to outside parties might be met with legal charges of culpable negligence. Corporations that wish to provide information to customers, subscribers, partners in a joint enterprise, fellow state, local, or federal agencies, or for publication, should be more strictly controlled.

Fourth, to comply with some of the regulations that have been proposed would require a vast amount of work and expense. It is important that society's eventual legal framework for privacy protection be one that lends itself to the maximum automation. The computer must keep check on the computers. Only in this way can the ultimate system be workable.

The main proposals for new laws require that data banks containing information on individuals should be registered and their purpose clearly stated. They should then hold only that data which is relevant to the stated purpose. Facts, not opinions, should be stored. A log should be maintained of all interrogations of these data banks.

The public should have the right to inspect the records which the computers keep about them and to take issue if they believe the data to be incorrect, out of date, misleading, or irrelevant. The register of data banks should be open to inspection by the public and particularly by the press, so that the public can be informed what *types* of information are stored.

It would be a criminal offense to avoid registering a data bank or to do so falsely, and a negligent data bank operator should be liable for damages.

The law is an evolving structure, and the bills that have been proposed could provide a foundation on which to build. Human flexibility is important in the laws. No doubt as the new laws come into operation, unforeseen loopholes will become apparant and can be sewn up.

It is desirable to make these safeguards as automated as possible; otherwise, a massive amount of work could result. If a registrar, for example, is likely to inspect the interrogation log kept by computer centers, then this log should be in a machine-processable form. The format of the items it keeps should be a format specified by the office of the registrar. It should become a requirement that such logs can be *transmitted* to the registrar's computer if required. The registrar's staff can then inspect them on their screens and ask the computer such questions as "On what occasion did so-and-so ask for information?" or "How often has this file been used for this purpose?"

A society with a high level of automation must eventually frame its laws and safeguards in such a way that computers can police the actions of other computers. Over a period of time appropriate checks and balances will grow up in this fashion, so that we can develop the type of environment we want to live in. Computers can protect us from harassment by other computers.

Dr. Emanuel R. Piore, formerly chief scientist of IBM, summed up the situation as follows.[5]

> None of these facts and prospects should obscure the plain truth that in the end, preservation of privacy rests not with machines but with men. The effectiveness of all protective measures, however sophisticated they may become, will still depend upon people: operators, service personnel, supervising officers, and all those who decide what information to put into a computer and how to use it.

Chapter Twenty-Two

WIRED WORLD

Could we not leave these remote villagers alone as they have been for centuries? That is not the way of the satellite age. A new generation of young people will seek their fortunes. Some will rage against the inequity. Some will invent, create, and be intoxicated with the opportunities of this bubbling, racing, heady, turbulent planet.

When traveling in foreign communities it is interesting to look at the faces of the people and try to assess their level of happiness or contentment with their lot. Clearly this can only be a subjective feeling, and sociologists cannot convert it to statistics. It is interesting to do because in some communities there is a feeling of great discontent, unhappiness, restlessness. The people want to change but are trapped. There are dangerous frustrations. However, in many of the villages which we pity from afar when we read the statistics, there is a feeling of content. The per capita income, easily measurable, is $295 per year. Appalling. But the faces of the people are happy—a fact you cannot measure, quantify, or put on a chart. We tend to know only those facts which are measurable or are respectable in academic papers. If you travel away from the tourist tracks, try to ask yourself, Why are these people happy and these people not? Why, in the remote parts of Nepal where you could not even take a jeep and where the income is among the lowest in the world, are the villagers humming with contentment, their faces unquestionably happier than the faces on the streets of New York?

It has struck me time and again that the contented faces are in areas where there is no electronic communications. The people are happy with their lot because they do not know of any different way of life. They know their place in the hierarchy of their village, and their aspirations are to grow fat vegetables, to see their children grow up, and to be accepted in their community. They do not know that the United Nations classes them as "developing." They live in well integrated communities where the patterns of life have been honed over the centuries.

Much of the world's population lives in such villages. In the next ten or twenty years a television set is going to be introduced into many of these communities. Most people in densely populated parts of the world like India, Africa, and China do not see television today. Most do not even have radios. Furthermore developing nations of the world have to use foreign television for about half their programs, because they cannot afford the production costs, and most of the imports come from the United States. Guatemala imports 89 percent of its television, mostly American.

A communications satellite is far away in space, so that almost half the earth is in its view. A satellite channel from London to Rio, or Toronto to Johannesburg, *ought* to cost the same as a short-distance one. The facilities we have described can therefore be linked worldwide. Mail and messages, television and talk, pass around the world in a fraction of a second. It is as easy to dial Rome from my New York office as to dial across the street. We speak of a wired society; a wired world is not far behind.

In 1974 NASA launched a satellite with an antenna like a thirty-foot umbrella which opened in space.[1] This umbrella could beam television down to relatively low-cost receivers. The satellite was moved around the equatorial orbit to India, where television was transmitted to several thousand villages. Prior to this only a small area around New Delhi, with a total of about ten thousand sets, had received television. The antennas for picking up the satellite television were made locally with chicken wire. Now India and other developing areas plan to have their own satellite. In a decade or so, hundreds of millions of people now isolated from communications will join the world's television-watching hordes.

There are good reasons why developing nations want to use television. As we mentioned earlier, this immensely powerful medium can enlighten, educate, spread literacy, and spread better farming methods. It is probably the only way of training the work force needed in countries like Iran. Medical assistance can be brought to areas where doctors or specialists are in short supply. Experiments in telemedicine have been highly successful, many of them performed with the same satellite that brought television to India. The satellite provided two-way video channels between doctors and patients. Satellites can transport other forms of expertise also, making available crop disease experts, consulting engineers, or world authorities at remote locations where their special skills are needed.

WORLD TELEVISION

Radio and television have done much to homogenize culture.

In England fifty years ago there was a wide diversity of

regional accents. Some were so strong that they were difficult to understand. An expert like Professor Higgins in Shaw's *Pygmalion* could identify both the home region and social class of a speaker. A young person from the provinces seeking his fortune in London strove hard to change his accent. Today the strong accents and provincial volcabulary are largely gone. Nearly everyone speaks with the range of accents used on radio and television.

In the 1930s, 1940s, and 1950s most of the world listened to the same pop music—Gershwin, Dixieland jazz, Louis Armstrong, the tunes Fred Astaire danced to, then Frank Sinatra, Elvis Presley, Rock and Roll; it was almost all American, and it was carried around the world by radio. If you went to a Saturday night hop in Costa Rica, Brazil, or Hong Kong, most of the dances would be to American music. Few Americans realized it, but they had set the whole world to dancing their tunes. Today these countries have more local music, as local recording and broadcasting studios have grown up.

Television is following a similar pattern. The total foreign sale of U.S. television is 150,000 program hours.[2] Britain is second with sales of 20,000 hours. France and Germany are third and fourth. Except for these four countries television exports are very small. There is a small exchange of programs within the iron curtain countries and within Spanish-speaking Latin America. Russia, China, Japan, and the United States import little television, but much of the rest of the world imports one third to one half of what it views—mostly from America. Most countries buy all or most of their series programs and feature films from abroad.

The bulk of television exports from the United States are feature films, comedy shows, and series like *Ironside, The Lucy Show, Bonanza, Perry Mason,* and (particularly popular) the 514 episodes of *Peyton Place.* Several U.S. series contain more than 500 episodes and fill several hundred hours of programming. Many of these programs are dubbed in five or six languages. The United States exports very few educational, current affairs, or serious programs.

As is often the case, this use of the medium is not a result of conscious policy planning but of accident, resulting from the nature of the technology and production costs. It costs over $100,000 to make one episode of *Peyton Place*. A country like Finland buys one episode for about $500, the cost depending on the number of receiving sets. Most countries have spent much money on the transmitting facilities and feel obliged to keep the channels full, but they do not have the money to produce enough films and entertainment. Current affairs and talk programs, on the other hand, are cheap to produce. There are plenty of local people, especially university professors, only too eager to hear themselves on television, and you need not pay them more than a token amount.

Television is thus doing little as yet to enhance world education, spread the best of human culture, and foster understanding between peoples. However, world television is in its infancy and is rapidly developing. Some countries make excellent documentaries and current affairs programs that would do much to help international understanding if they were seen by as many countries as *Peyton Place* is. As the numbers of channels increase and the cost of international transmission drops, it is likely that such programs will find larger distribution.

WORLD NEWS

World news distribution also tends to be a one-way street. International news film is expensive to produce. A small country cannot afford to have its own film crews covering world events. It therefore has to buy film footage from the richer or larger nations or from international agencies, which are almost all American or European. The circulation of international news magazines follows the same pattern: first, America with *Time* and *Newsweek*, and then, much smaller in circulation, magazines from England, France, and Germany.

In the 1970s there has been growing concern in many nations about this one-sided distribution of news, current affairs information, and television shows. It is commonly referred to in the recipient nations as "cultural imperialism."

The constitution of UNESCO (the educational and cultural arm of the United Nations) calls for "free flow of information" between nations, but in the 1970s UNESCO has been making declarations opposed to *laissez faire* in communications on the grounds that most of the information flows from a handful of nations. Instead, they have advocated reducing the lack of balance in international communications.

The Seventeenth General Conference of UNESCO in 1972 declared that transmission of television programs by satellite to the population of another country should be condemned unless there is explicit agreement with the recipient country. The Soviet Union strongly advocated this declaration.

In 1976 UNESCO advocated that Third World countries band together to set up a collective government-backed news agency designed to correct the "serious imbalances" in news flow in and out of the Third World. It advocated the creation of "national communications councils" to draw up "guidelines" for the press. It proclaimed "the need for state investment in the mass communications sector in accordance with that sector's priorities and responsibilities within overall development planning."[3] The Western press reacted with great alarm to the proclamation, assuming that the Third World news service would be a powerful means of controlling the news according to the needs of authoritarian governments. Many countries today *do* have government control of the news. Only 19.8 percent of the world's population lives in countries with a free unfettered press.[4]

MULTIPLE CULTURES

It seems probable, then, that technology will not yet bring equal communication channels to all people. It will not spread the culture of different peoples uniformly. It will not completely open windows between nations. The wired world will have multiple cultures, and they will be woven into global patterns by telecommunications and agreements between nations. One cultural thread will be that of the free-press nations, who will increase their interchange of news and current affairs programs. Other threads will be entertainment from America, television drama from Britain, and films from France.

When we speak of great civilizations today, we refer to *geographically bounded* regions such as ancient Greece, eighteenth-century France, modern America. It is possible that after a few decades of global communications, "civilizations" will be worldwide threads. A postindustrial Western civilization may link the world's English-speaking peoples. Chinese, Russian, Latin American, or Third World civilizations may also become global threads but with fundamentally different values and cultures. New global influences will grow up with different music, art, values, and viewpoints.

In cosmopolitan cities the different threads will merge. Enthusiasts of one culture will tune in to that global thread, for example watching Japanese television, going to Kabuki theater, eating Japanese food, and absorbing Shinto religion.

GLOBAL CORPORATIONS

Particularly strong threads in this global texture will be the multinational corporations.

To operate effectively, a corporation must have good information sources and excellent communications. As world telecommunications improve, corporations will be able to tie together their national operations with corporate communication networks.

Worldwide travel is more onerous than national travel, so there will possibly be a greater incentive to devise effective person-to-person telecommunications globally than nationally if the cost becomes similar. The leased-line networks, electronic mail, video-conferencing, voicegram systems, and computer links will become an essential facility of global corporations, vital to their fast, efficient functioning. As we have said, the expenditure on air travel is very high in some worldwide corporations. Some of this travel is unavoidable, but for a large part of it telecommunications could form a substitute if effective video, facsimile, and computer facilities were available, all capable of being hooked into conference calls.

Funds can be electronically transferred from one country to another and switched to different currencies. Ships are becoming linked into the corporate telecommunications networks by means of satellite systems. Optimal fleet scheduling is a computer application which can save much money. Inventories can be maintained and resources moved around on a worldwide basis.

Multinational corporate networks, like national ones before them, will raise severe arguments about centralization and decentralization. The local staff should handle local situations, but there are some functions that arguably might benefit from centralization—bulk buying, inventory control, product design, capital management, some computer services, guidance in software development, and so on. Video conference rooms and computerized information systems increase the degree to which head-office executives might attempt to guide or interfere with management in other countries.

The American military has superb worldwide communications. The effect of such fast and well integrated information flow is that high-level commanders often participate in decisions that formerly would have been made at some lower level: The buck is passed upward at great speed. Often the Pentagon chiefs are involved in what were previously field decisions. In the Vietnam war the press reported how President Johnson became involved in such command decisions; for example, he personally reviewed each day's bombing targets. Field commanders, it is sometimes reported, resent such interference from on high, but the technology now makes it possible to bring the best brains together on critical situations. To an increasing degree, command will be exercised from distant war rooms and will involve persons and computers in different locations connected via terminal screens. The same will become true in multinational corporations.

Global corporations could do more to spread international understanding than any other force. They provide employment, spread technology, foster equality of aspirations, and train large numbers of skilled people. There is concern that the multinational corporations will become too powerful and that international laws are not adequate to control them. The largest ones operate with budgets larger than the national budgets of small countries, so it is not surprising that some governments fear them.

Most people still feel stronger emotional ties to their country than they do to their corporation. Volkswagen employees in Germany feel more strongly that they are Germans than that they are Volkswagen people. But patriotism is declining and may decline more with decades of advanced global communications. Some people will feel more loyalty to their global cultural thread than to their country. An English computer specialist with IBM feels more in common with a Japanese IBMer than with an English poet. American and Russian astronauts together exhibit understanding and loyalty that transcends national differences. The closest friends of a professional person or a university professor are often not neighbors but persons with common interests, who might live anywhere. Such people have a network of colleagues, and the network can be worldwide, linked by telephone, data networks, and occasionally jet travel. Associates of the same cultural thread have periodic world rallies with excuses for their firm or university to pay the fare. When facilities have improved, they will see each other on the corporate or college teleconferencing screens.

Some Europeans now claim to be Europeans first and Germans or Belgians second. If the technology described in this book had come forty years earlier, the British Commonwealth might never have disintegrated. We hear corporate executives make statements such as "We do not think of ourselves as a Dutch company but as an international company with a head office in Holland."

The shape of cultural patterns is often determined more by money than by aesthetic or abstract values. The imperative to maximize profits will increasingly be an imperative to market internationally and hence design products for international markets. The better the worldwide flow of information, the more practical this will be. The economics of television production will increasingly depend on the world market rather than national markets for the program. Coproductions between countries will become more common, and multinational organizations will produce entertainment, educational, and information services, and new films.

For reasons probably related to the spread of global communications and jet travel, many multinational corporations are growing faster than equivalent national ones. They include manufacturing corporations, conglomerates, mining firms, news agencies, and service corporations such as banking. They are coming into existence by a global concentration process. We see international mergers such as those in Western Europe to meet the challenge of America and Japan. Binational and trinational mergers such as the Dunlop–Pirelli merger occur, and finally we have multinationalism, with multinational boards of directors and shareholders. New entrepreneurs are making fortunes by starting out in a multinational fashion. Tiny corporations as well as large ones are succeeding by using multinational market planning. In another pattern, large national corporations continuously expand their operations, setting up subsidiaries around the world.

The advantages of multinationalism are many. Such corporations have a vast worldwide market. They diversify their risks. A slump in Europe does not necessarily coincide with a slump in Japan or Brazil or the Arab world. They can usually raise money more easily and manage currencies more adroitly. And above all, they have large international markets.

If you drive around the industrial suburbs of Sao Paulo, the world's fastest growing city, you will recognize the names of most of the factories. They are household words—Kodak, Pepsi Cola, Volkswagen, Olivetti, Shell. . . . These are the large multinationals. There are also small ones concentrating on highly specialized markets.

In ancient Greece there were city-states. Later there were nation-states, which merged into larger groupings. A pattern of the satellite age may be something rather like corporation-states, employing hundreds of thousands of people worldwide, linked together by global video, voice, and data links, moving funds around the world electronically. Countering them will be multinational unions capable of organizing multinational strikes. The corporation-states will compete fiercely, but they are unlikely to go to war. The loser of an economic war is better off than the winner of a war with modern weapons.

SATELLITES OF MANY TONS

It is difficult to escape the conclusion that today's communications satellites are but an elemental beginning of a technology with tremendous growth ahead of it. One day readers of this book will look back at WESTAR, the first U.S. domestic satellite, with the amused but admiring sense of history that we have when we look back at Explorer I, America's first grapefruit-sized response to Sputnik. WESTAR weighs only 300 pounds. At some time during the 1980s it will become economically reasonable to take hardware weighing many tons into geosynchronous orbit.

A new era in space will begin in 1980, when the space shuttle is operating. We will look back at today's throw-away rockets as an incredible waste that was necessary in order to make the first steps into space. The space shuttle will ferry cargo into space and then return to earth and land at an airport, ready for its next trip. The satellites used for world communications today are less than eight feet in diameter. The space shuttle has a cargo bay sixty feet in length and will make it possible to join together many modules in space, if desirable, to form very large satellites. Today's satellites generate only enough electricity to light a few light bulbs. Larger satellites will generate one hundred or one thousand times as much power. We can build immensely powerful satellites which can beam down to our rooftops as much information as society can use.

A vast industry will grow up, placing massive hardware in orbit in a ring around the earth 22,300 miles above the Equator. Eventually, it will become economical to have service vehicles in geosynchronous orbit, repairing, refueling, or assisting in the deployment of the satellite equipment. It has been suggested that geosynchronous orbit will become the right place for certain new manufacturing processes which need the intense vacuum of outer space and perhaps the absence of gravity. The vacuum of geosynchronous orbit is more perfect than in low orbit, and the solar power generation would not be interrupted each day by the earth's shadow. Being stationary above the earth, the production process would be constantly linked to terrestrial computers, control rooms and video monitors. One can imagine future solid-state logic circuitry or computer memory being fabricated in the utter purity of space with microscopic components thousands of layers deep being deposited on silicon. It is fascinating to read the original 1945 article in *Wireless World* in which Arthur C. Clarke first proposed communications satellites.[5] He argued that it would be prohibitively expensive to build a terrestrial network for trunking television or wideband signals:

> The service area of a television station, even on a very good site, is only about a hundred miles across. To cover a small country would require a network of transmitters connected by coaxial lines, wave guides, or VHF relay links. A recent theoretical study has shown that such a system would require repeaters at intervals of fifty miles or less. A system of this kind could provide television coverage, at a very considerable cost, over a whole of a small country. It would be out of the question to provide a large continent with such a service, and only the main centers of population could be included in the network.

Clarke suggested that satellites would eventually be far less expensive than terrestrial links, and most of the details he gave of satellites were surprisingly accurate. However, he greatly underestimated the money that would be spent. The terrestrial networks were in place before satellites could be used. Nevertheless, the simple logic of his argument is valid. Satellites, now that they are practical, are vastly cheaper than continental networks of coaxial cable, microwave relays, and waveguide systems. Hence the wiring of much of the world, which does not yet have the vast investments in terrestrial links, may come from satellite systems with the new glass fiber trunks in corridors of high traffic density.

Certain developing nations that do not have the advanced terrestrial networks of Europe or America are planning satellites. Some countries which had major difficulties in deploying telephone cables, because the villagers would steal the wire to tie up their goats, are now planning satellite antennas on the roofs of telephone exchanges. It seems probable that some of these nations will have more advanced satellite systems than Europe in ten years' time, and their capital development costs for telecommunications will be less.

UNEVEN SPREAD

The power of telecommunications will not spread evenly around the world. Today most of the world's persons do not have a telephone. North America, Western Europe, and Japan have five sixths of the world's telephones but only one sixth of the world's population. The United States has 66 telephones per 100 people. Most developing nations have less than one telephone per 100 people, and almost all of these are in industry or government. Many areas of the world are barely penetrated by television. There are fewer television sets in the whole of black Africa than in Hong Kong.

The benefits that new telecommunications can bring to a society will accrue mainly to the richer societies. A wealthy developing society can develop far faster than ever before in history by employing the new media, which bring information sources and educational television.

Ten years ago Iran was one of the most primitive of nations; as we commented earlier, Iran plans to achieve a GNP comparable to Western Europe in 20 years and this can be done only if a large number of people are educated and trained for new skills at a very rapid rate. Consequently a satellite has been designed which could serve two thirds to three quarters of the Iranian population. Educational satellites could prove an extraordinary investment for other developing nations.

DEVELOPING NATIONS

Indonesia has a far lower telecommunications penetration than most developing nations. It has only one telephone to every 500 people. Yet Indonesia was the third nation after Canada and the United States to launch a geosynchronous satellite for its own domestic telecommunications. The satellite was the same size and type as the first North American satellites.[6] An Indonesian Prime Minister in the fourteenth century, Gajah Mada, vowed not to eat *palapa,* a great delicacy at that time, until all the peoples of the country were united. To commemorate his sacrifice the Indonesians call their satellite *Palapa I.*

The Indonesian satellite illustrates an important principle. In the developing world, communications facilities are built in a different sequence from that in the West. The West had railroads, then roads, and later the airplane. In Brazil many towns can be reached only by air; later, roads are built leading to them, and later, if at all, railroads. Similarly, satellites will bring the links of culture and commerce to towns without telephone trunks. Data radio will be available to locations that cannot send data by telephone. Cable television will reach some homes before the telephone.

Rather than duplicating the facilities of the West, developing nations should use the new technology in new ways related to their own special problems and opportunities.

Not only should the long-distance links be of a different technology in the developing nations, but also the local distribution may be different. Such nations do not need a vast acreage of the radio spectrum to be allocated to television. That leaves plenty of spectrum free for allocation to citizens' band radio, radio telephones, packet radio, paging, and broadcast data. These services are relatively more valuable in areas where telephone cabling is not as ubiquitous and as high in quality as in the major industrial countries. A businessperson may find a radio telephone invaluable in places where conventional telephones are poor or nonexistent. Packet radio units may be attached to computer terminals instead of the cabling we use—and they could cost less. They may be linked to communications satellite relays, as in the University of Hawaii system, thus permitting the spread of computer usage in areas where telephone links are bad.

The best approach to providing telecommunications service in developing nations is not to copy the traditional technology of the West—the copper telephone cables and electromechanical switching. These represent the technology of an earlier era when copper was cheap and electronics expensive. Instead, developing nations should use new technology satellites and radio, computers for switching, community telephones rather than individual phones, and data networks for industry and government.

TRAUMATIC CHANGE

As we illustrated earlier, the introduction of new communications technology brings major upheavals in society. To take a primitive society and rapidly introduce advanced media is a formula for chaos. South Africa delayed the introduction of television until the mid-1970s and then rigorously controlled it. It is doubtful whether South Africa could withstand the full impact of Western television. In some areas of the Moslem world movies, dancing, and alcohol are banned, and television fades out when a man and woman approach to kiss. A cowboy slamming down a dollar on a bar asks for lemonade on the dubbed soundtrack. In most of the world, however, media censorship has not fully worked. It is almost impossible to reedit all Western shows. With modern technology, banned radio programs get through, and earlier attempts to jam radio stations have usually been abandoned.

The wired world, then, will have few hiding places. We could say, Why interfere with the agricultural communities? To understand India and Africa you must understand the villages. Their social strata have been unchanged for centuries. The aspirations of their people lie within the village community. They have little or no knowledge of the world beyond the village. Television will drastically change the aspirations of the developing world. Hundreds of millions of people will realize that opportunities exist elsewhere. They will realize what money can buy. Primitive people will be hammered with the most sophisticated and cunningly persuasive advertisements—all the skill of Madison Avenue—because this is how multinational corporations will maximize their profits.

Could not we let well enough alone and leave these happy villagers at peace as they have been for centuries? That is not the way of the satellite age. By the year 2000 the planet will have 6 billion people bombarded with the emotional intensity of video media. Villagers will move to the cities. A new generation of young people will seek their fortunes. Some will rage against the inequity. Some will invent, create, and be intoxicated with the opportunities of this bubbling, racing, heady, turbulent planet.

Until now the mind of man has manifested itself in a static fashion in isolated groups. What sort of current will be generated when the world billions are suddenly wired together?

Chapter Twenty-Three

WHO PAYS?

U. S. Federal spending on highways (not counting state and city spending) exceeded $70 billion in one ten-year period. Now we need electronic highways. A similar expenditure on communications satellite facilities would work miracles.

Many, but not all, of the services we have described in this book would be profitable. Private corporations could provide them, raising capital in the normal ways. In countries that frown upon private enterprise in telecommunications, government departments could provide them without subsidy if they were run efficiently

Some of the services we have described could be operated only with subsidies or would be unlikely to start without government seed capital. If there were no government intervention, many of the services would go only to densely populated areas or to higher-income groups, and this might be socially undesirable or disruptive. Education and medical services, news, and audience-response systems for political or community issues, for example, ought to be available to all members of society, not just certain cross sections.

Some communications services are subsidized today even in the most capitalistic countries. Noncommercial broadcasting is paid for by governments. National mail services do not pay their own way. If telephone service were run as a strictly business operation, the service in rural areas would be very expensive or nonexistent. Subscribers on high-density cable routes subsidize those in sparsely populated areas. Businesses pay more than homes for telephones in the United States, although their connection cost per telephone is usually lower.

However, such subsidies are small compared with the subsidies for road building, which have amounted to many billions of dollars per year in the United States. Airports and mass transportation are also massively subsidized.

There is much hidden taxation behind a country's communications facilities. It is designed to bind the country together socially and provide an infrastructure which permits commerce and government to operate more efficiently.

NEW SUBSIDIES?

There are several arguments for new forms of telecommunications subsidization:

1. Satellites, glass fiber trunks, and other new technology make possible excellent education and medical

services. However, the facilities needed may not be provided by private industry because the business risks are too high or the business justification insufficiently attractive.

2. Better news media and audience-response television could have a major effect on democratic government.
3. There are major economies of scale in satellite design. Very large satellites would permit low-cost earth stations and cheap channels of high capacity. They could be used either for broadcasting or two-way communication. It is doubtful whether any one corporation will launch the satellites that could benefit society so greatly. Judging by its track record, the best organization to design and launch America's satellites is NASA.
4. High-capacity cables into homes have many potential uses and will eventually play a major role in society. The cable television companies are installing them at a rapid rate in North America. They are being installed in the most profitable areas and, if left unsubsidized, will never reach the less profitable areas. Some of the low-profit areas are those most badly in need of the education, medical, and other services which the cables (and later optical fibers) can bring.
5. Television is the most powerful communications medium in history. Many of its potential benefits to society are not being exploited.

ADVERTISING

About 2 percent of the American gross national product is spent on advertising—a gigantic sum. In one sense it is used inefficiently. The American Association of Advertising Agencies has estimated that each day 1600 advertisements are aimed at an individual, but of these 80 are noticed and only 12 provoke some reaction. It is rather like fighting a battle by firing 1600 bullets in all directions at random. Only 80 reach the enemy, and only 12 hit anyone. Worse, the effectiveness ratio is going down. The public, bombarded with increasing quantities of messages and information, is overloaded. It has developed a protective shield to screen out the unwanted messages, and is learning to be cynical—to disbelieve.

To be more effective, the fire needs to be more concentrated, more selective. Some of the new news and information services we have described would permit this. If there were many channels of television, there could be highly specialized programs that appeal to more specialized consumer groups. *Scientific American* has a smaller circulation than many magazines, but its advertising is more effective because it is aimed at a narrower target. The same could apply to television. Interactive facilities, in which cable television subscribers do shopping or catalog searching from home, is much more narrowly targeted, and new types of advertising for this medium will be highly effective. Just as advertising financed the great U.S. television boom of the 1960s, so it could finance some of the new facilities. Doctors could watch commercials for new drugs on their office screens, farmers would have facilities provided by fertilizer companies available from their data terminals, and so on.

It could be valuable to make advertising pay for noncommercial services as well. A tax on television advertising could be used to finance a news network, for example. Audience-response facilities could be designed for both commercial and noncommercial use.

CROSS-SUBSIDIES

The driving force which has brought about most improvements in the *use* of technology comes from the entrepreneur and the innovative private corporation. The entrepreneur seems essential for progress in usage of telecommunications. Most telecommunications organizations which are monopolies have been remarkably backward in introducing new services to the public. However, there is a basic problem in allowing the entrepreneur complete freedom in telecommunications: Competition in the marketplace will bring innovations that permit some of society to have excellent service while more remote or expensive-to-serve segments will be left either unserved or served only at very high prices.

There are several ways to react to this problem:

1. *Do nothing.* Allow different facilities to develop in the

cities from those in the country; some of the people who want cable television or electronic library service will move to areas that have them. The trouble with this is that poor or rural people will be left without the new media and educational facilities.

2. *Let the government do everything.* In many countries the government operates all telecommunications facilities. The trouble with this is that government bodies in most countries have a poor track record for innovation and customer service. Some of them can barely cope with telephone service, let alone what we have described in this book.
3. *Have private industry monopolies operating under regulations.* The corporations have profit motivation but must cross-subsidize, letting the profitable areas help pay for service in the nonprofitable areas. This is the pattern of the U.S. telephone industry, which has provided good but not innovative service. The injection of a small shot of competition in the 1970s resulted in a major increase in motivation. There are few genuine "natural monopolies," and it seems desirable to keep the monopolistic segments of telecommunications to a minimum.
4. *Use combinations of government subsidies, a government operation, and private industry.* For example, NASA might launch a satellite, while the earth stations and ground systems might be designed and operated by private industry. Some of the industry would be made to cross-subsidize to ensure widespread use of the facility.

There are problems with all of these solutions. Those involving regulation of private industry result in bitter squabbling and lobbying about the regulations. All innovation and change in regulation is seen by some segment of industry as potentially harmful to its profits.

THE EFFECTS OF REGULATION

Telecommunications will probably always be regulated to a greater or lesser extent. The principle of regulation should be "maximize the benefits to the users." This has been far from the case with much regulation. The principle of U.S. telephone company regulation is to control the *rate of return* that they earn. This does not maximize user benefits; in fact, sometimes exactly the opposite has been the case. Rate-of-return regulation has tended to encourage telephone companies to have low depreciation and plant of large capital value. Most telecommunications plant is depreciated over forty years. This ensures that equipment will be used long after it has become obsolete and that accountants discourage technical innovation which will make the equipment obsolete. In the computer industry it is inconceivable to design or install a computer for a forty-year lifetime. Yet that is common in telecommunications, where technology is changing as fast as in computing. The users suffer, with old equipment at high prices and no new types of services.

Fortunately, telephone usage has been expanding rapidly for several decades, so that new technology has been installed to handle the growing traffic. The volume of conventional *telephone* usage may not grow fast in the United States in the future. Most people in North America have telephones and already make most of the calls they want. If telephone volumes do not expand, regulation which results in long equipment lifetimes will tend to cause stagnation.

The rate-of-return formula also encourages low maintenance expenditures. Thus not only do we have obsolete plant, but it is often ill-maintained. Some of the major U.S. cities have had serious problems with their telephone service because of this and because of insufficient attention to the new user class, who transmit data. Under the streets of large cities there are old-fashioned cables in great quantities, and many of them are in poor condition, rotted by old age and by the liquid and gases under the streets. Government regulation gives telephone companies little encouragement to replace them, but they are storing up future problems for subscribers.

The worst aspect of the rate-of-return regulation is that it tends to discourage projects which could bring a massive saving in capital equipment costs, as could the use of large telephone company satellites today. The domestic satellite issue is politically explosive, because satellites could bypass the established telephone trunks, carrying a large nation's toll traffic at much lower cost. However, satellites are the key to many of the socially valuable innovations we have discussed in this book. Government action ought to be encouraging the development of large communications satellites, not discouraging it. It is tragic that the talents and experience of NASA are not being used for this purpose.

Many of the advantages of new telecommunications that we have stressed arise from providing *rural* communities with the same facilities as the cities. However, it would be very expensive to take the new high-capacity cables to remote communities. The economies of today's Picturephone service, for example, are such that nonurban areas would be the last to receive it. The answer to this seems to be transmission from large satellites which bypass traditional terrestrial links. Thus, if government action or subsidies are used to wire a rural society, it should be done not as an extension of existing telephone company technology but as an entirely new departure.

Perhaps the worst form of regulation is that which artificially attempts to preserve current ways of operating even though they are obsolete. This often happens, because organizations with obsolete facilities attempt to use the regulatory authorities to protect themselves from competition. Conventional postal services, for example, are endangered by electronic mail. A bad form of regulation would be to prevent the new competition from operating. Equally harmful would be to give massive subsidies to the old service to make it artificially price-competitive. Humankind could benefit greatly from certain technologies that may never be developed if they are impeded with negative regulations and subsidies.

TELEPHONE AND TELEVISION

Another problem of regulation is the relation between the telephone and television companies. They developed separately, and regulation kept them separate. Telephone companies in North America cannot build or own CATV (cable television) cables, and the television companies cannot give telephone service. Now, however, the television set appears to be the ideal screen for a data terminal in the home. Many socially valuable services require both a picture-carrying cable into the home and the capability of the telephone network to switch circuits to any part of the country. More significant, telephone companies (particularly AT&T) have developed optical fibers, which are going to be the ideal transmission medium for carrying telephone, television, and other signals into the home. When this is the best technology, should telephone and television be kept separate?

SYSTEMS ENGINEERING

Probably the most cost-effective solution would be a massive piece of systems engineering, like that for the moon landing, in which all the telecommunications requirements of society are integrated into one long-range system development plan. Against this is the argument for competition and entrepreneurs. Europe in particular could benefit enormously from a Common Market NASA-like organization, which would integrate European facilities for communications satellites, computer networks, television, electronic mail, and telephone. If Europe does not do something like this, it will not have the communications infrastructure needed to help compete with Japan and North America in the 1980s.

However it is accomplished, the building of the facilities we have described will have a major effect on society. The full benefits will arrive earlier if government funding is used. Massive government money was used for other high-cost technologies including jet airports, power stations, nuclear research, and the moon shot. It is even true with automobiles, because one must count the cost of the highways. A car without a road is like a Picturephone set without communication lines. U.S. Federal spending on highways (not counting state and city spending) exceeded $70 billion in one ten-year period. Now we need electronic highways. A similar expenditure on communications satellite facilities would work miracles (a spectacular satellite costs less than $70 *million*).

Some governments will probably spend the money. Others will lack the foresight.

To return to the theme of the first chapter, we believe that the correct choice of technological evolution can make our life on this planet better and richer. Wrong use of technology, or failure to apply new technologies, could be disastrous.

There is a danger that politics, lobbying, monopolistic sloth, regulatory ignorance, or vested interests will rob us of part of the riches that the technology could bring. It behooves our politicians and regulators to understand fully the many facets of possible future developments in telecommunications. However, it seems sadly possible (as with some other public uses of technology) that through lack of such understanding, we will fail to make full use of the new opportunities for enriching our world.

REFERENCES

Chapter 3: Medical Facilities

1. Raymond Murphy et al., *Observations on the Feasibility of Telediagnosis Based on 1000 Patient Transactions* (Boston: Massachusetts General Hospital, 1971).
2. Kenneth T. Bird et al., *Teleconsultation, A New Health Information Exchange Program* (Boston: Massachusetts General Hospital, 1971).
3. From *Five Patients,* by Michael Crichton. Copyright © 1970 by Centesis Corporation. Reprinted by permission of Alfred A. Knopf, Inc. International Creative Management and Jonathan Cape Ltd.
4. W. I. Card et al., *On-Line Interrogation of Hospital Patients by a Time-Sharing Terminal with Computer/Consultant Comparison Analysis.* Proceedings of the IEE conference on man–computer interaction, held at the National Physical Laboratory, Teddington, England, September 1970, Institution of Electrical Engineers conference publication, 1970, 68, 145.
5. This statistic is for 1969. Federal physicians are those in military service, veterans administration hospitals, National Institutes of Health, etc. Source: J. N. Haug and G. A. Roback, *Distribution of Physicians, Hospitals, and Hospital Beds in the U.S.* American Medical Association, 1970.
6. The following chart (from the report listed in reference 5) shows how the distribution of doctors varies with average income in the United States.

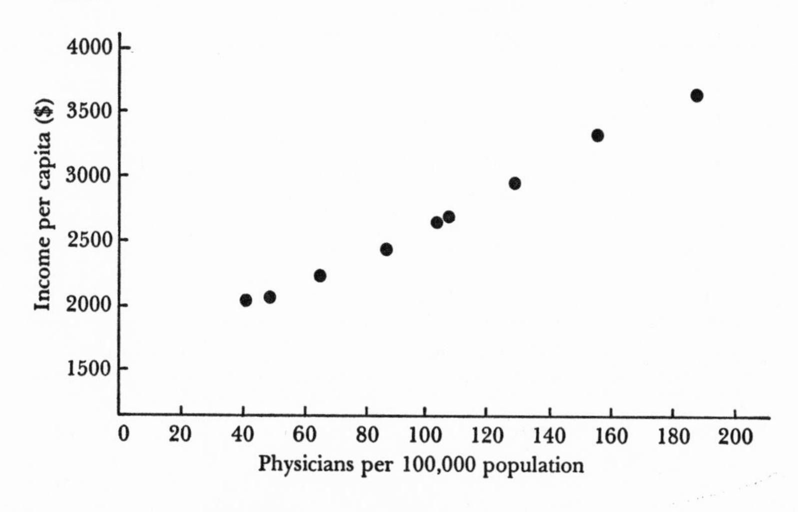

7. Details available from NASA, Washington, D.C.

Chapter 4: Widely Different Requirements

1. A combination of optical fibers, millimeterwave and microwave radio, and satellite transmission, all with digital multiplexing. Described in James Martin, *Future Developments in Telecommunications,* 2nd ed. (Englewood Cliffs, N.J.: Prentice-Hall, 1977).
2. Aldous Huxley, *Brave New World,* Chatto and Windus 1932, Penguin Books paperback 1974.
3. North America has four main levels of digital transmission, standards originated by the Bell System:
   T1 carrier: 1.544 million bits per second
   T2 carrier: 6.312 million bits per second
   T3 carrier: 46 million bits per second
   T4 carrier: 274 million bits per second
   The latter bit rate is carried by the tubes in a coaxial cable, the channels of a DR18 millimeterwave radio system, the channels of a WT4 waveguide, experimental optical fibers, and the transponders of a suggested satellite system.
4. Detailed examples of this are given in James Martin, *Telecommunications and the Computer,* 2nd ed. (Englewood Cliffs, N.J.: Prentice-Hall, 1976).

Chapter 5: New Uses of Television

1. Advertisement for Manhattan Cable Television.
2. Musak planned in 1975 to use the transponder of the WESTAR satellite to broadcast four high-fidelity music channels. A receiving unit with a four-foot antenna was demonstrated for receiving the music.

Chapter 6: News

1. Definition used by a multimillion-dollar media consultant firm, Frank N. Magid Associates, as the definition of "Action News."

Chapter 10: Instant Mail

1. The North American standard and world CCITT standard for digital telephone use channels of 64,000 bits per second. Of these 56,000 bits per second are used for speech and 8000 bits per second for signaling and control. *CCITT Recommendations* A.732 and A.733 (Geneva: International Telecommunications Union, 1973).
2. AT&T's T1 Carrier, widely used, transmits 1.544 million bits per second over copper wire pairs. AT&T's T2 Carrier carries 6.312 million bits per second over copper wire pairs.
3. AT&T's T4 Carrier operating over coaxial cables and other media transmits 274 million bits per second. This and the above are described in James Martin, *Future Developments in Telecommunications,* 2nd ed. (Englewood Cliffs, N.J.: Prentice-Hall, 1977).
4. H. Newton, "Communications Lines," *Business Communications Review* 6:5 (Sept.–Oct. 1976).
5. I. M. Nilles, F. R. Carlson, P. Gray, and G. Hanneman,

Telecommunications–Transportation Tradeoffs (Washington, D.C.: National Science Foundation, 1974).

Chapter 12: Radio Devices

1. AT&T cellular radio telephone technology, described in James Martin, *Future Developments in Telecommunications,* 2nd ed. (Englewood Cliffs, N.J.: Prentice-Hall, 1977), Chapter 12.
Motorola portable telephones. See *Future Developments in Telecommunications,* 2nd ed., Chapter 12.
3. Ceefax is described in *Future Developments in Telecommunications,* 2nd ed.
4. T. Johnson, "A New Read-Only Terminal: Your Television Set," *Data Communications, (May 1975).*
5. N. Abramson, "Another Alternative for Computer Communications," *AFIPS Conference Proceedings,* FJCC, 1970.
6. (a) L. Roberts, "Extension of Packet Switching to a Hand Held Personal Terminal, *AFIPS Conference Proceedings,* SJCC, 1972.
   (b) Collins Radio Technical Report 523–0602169–001C3L, *Packet Radio Communications* (Dallas: Collins Radio Group, Rockwell International, 1976).

Chapter 13: The Satellite Age

1. Comsat's first four satellites, EARLY BIRD, INTELSAT II, INTELSAT III, and INTELSAT IV, had an investment cost per voice channel per year of approximate $23,000, $11,000, $16,000, and $618, respectively.
2. Leroy C. Tillotson, "A Model of a Domestic Satellite Communication System," *Bell System Tech J.* (December 1968).
3. Described in the SBS filing to the FCC (Washington, D.C.: Federal Communications Commission, 1976).
4. P. E. Glaser, O. E. Maynard, J. Mackovcisk, Jr., and E. L. Ralph, *Feasibility Study of a Satellite Solar Power Station,* NASA Contractor Report CR-2357, Washington D.C., 1974.
5. "Satellite Solar Power Station: An Option for Power Generation," briefing before the Task Force on Energy of the Committee of Science and Astronautics, U.S. House of Representatives, 92nd Congress, Second Session, Vol. II (Washington, D.C.: U.S. Government Printing Office), 72–902–0.
6. Ivan Bekey and Harris Meyer, *1980–2000: Raising Our Sights for Advanced Space Systems* (Los Angeles: The Aerospace Corporation, 1976).
7. James Martin, *Communications Satellite Systems,* Prentice-Hall Inc., Englewood Cliffs, N.J., 1978.

Chapter 16: A Substitute for Gasoline

1. Statistics from *Newsweek,* 1975.
2. $1 billion could equip many corporations with new communications networks using the technology of Satellite Business Systems, the subsidiary of IBM, Comsat and Aetna.
3. Charles E. Lathey, *Telecommunications Substitution for Travel: An Energy*

Conservation Potential (Washington, D.C.: Office of Telecommunications, 1975).
4. Arthur Goldsmith, *Telecommunications–An Alternative to Travel* (speech) U. S. Department of Transportation, Washington, D.C., June 5, 1974.
5. Communications Studies Group, Joint Unit for Planning Research, *The Scope for Person-to-Person Telecommunications Systems in Government and Business* (London: University College, 1973).
6. *Communications News,* October 1974.
7. Richard C. Harkness, *Telecommunications Substitutes for Travel,* COM-74-10075, OT-SP-73-2, (Springfield, Va.: U.S. Department of Commerce, 1973).

Chapter 17: Industry

1. From a draft on an article by Harvey L. Poppel and Anthony G. Ward, "Time to Tame Telecommunications," Booz-Allen & Hamilton, New York, 1975.
2. The $10 figure is from a Booz-Allen & Hamilton study and is quoted in reference [1].
3. The cost of storing voice messages in a large on-line store such as the IBM 3850 would be about $.20 per year per message for messages coded with delta modulation, $.02 per year for messages coded with vocoder techniques, and $.0001 per year for messages composed with code-book words. Many messages would not be stored on-line for a year but for at most a few weeks. The cost of mass storage is dropping rapidly.
4. William E. Workman, "Which Color Washer Will They Choose?" *Computer Decisions* (December 1969).
5. The SBS (Satellite Business Systems) facilities are described in a filing to the U.S. Federal Communications Commission, 1976. The satellites operate at 12/14 GHz and are designed so that corporate locations will have equipment that can transmit directly via the satellite data, voice, facsimile, or video signals.

Chapter 19: Education

1. Figures taken from "There's a Computer in Your Future," *American Education* (November 1976).
2. Lawrence T. Backka and Bruce B. Lusignan, *Educational Television and Development in Iran, IEEE Transactions on Communications,* Vol. Com-24, No. 7 (July 1976).
3. Data from Pahlavi University, Shiraz, Iran, for 1975, quoted in the above article.
4. Satellite Business Systems FCC filing, describing satellites to be launched in 1979, Federal Communications Commission, Washington, D.C., 1976.

Chapter 20: 1984

1. George Orwell, *1984*, Harcourt, Brace & Co., Inc., New York 1949, now a Signet paperback.
Something approximating Orwell's telescreen could be built with

AT&T Picturephone technology or with an extension of CATV. To be really practicable it needs optical fiber local loops, which are likely to be installed in the 1980s or 1990s. Orwell did not foresee computers, although the first ones were in use.
2. Statistics on television viewing from A. C. Neilsen Company, New York.
3. The ten corporations controlling 78 percent of U.S. television in 1967 were: Procter and Gamble; Bristol-Myers; General Foods; R. J. Reynolds Industries,; American Home Products; Colgate-Palmolive; General Motors; Gillette; Sterling Drug; Lever Brothers.
4. Taken from a UNESCO report, *Television Traffic, A One-Way Street?* (Paris: Unesco Press, 1975).
5. Max Lerner, *America as a Civilization* (Touchstone-Clarion, 1967).

Chapter 21: The Human Goldfish

1. "The Computer and Invasion of Privacy," p. 37, House of Representatives, 89th Congress, 2nd Session. July 26, 27, and 28, 1966 (Washington, D.C.: U.S. Government Printing Office, 1966). (Subcommittee Chairman: Cornelius E. Gallagher, New Jersey.)
2. From James Martin and Adrian R. D. Norman, *The Computerized Society* (Englewood Cliffs, N.J.: Prentice-Hall, 1970).
3. James Martin, *Security, Accuracy, and Privacy in Computer Systems,* (Englewood Cliffs, N.J.: Prentice-Hall, 1973).
4. Three typical examples of legislation are:
   a. The United States Fair Credit Reporting Act of 1971.
   b. The United States Federal Privacy Bill of 1971.
   c. The British Data Surveillance Bill of 1969.
   The main items in these are summarized at the end of Chapter 39 and are given in Appendix C of James Martin, *Security, Accuracy, and Privacy in Computer Systems* (Englewood Cliffs, N.J.: Prentice-Hall, 1973).
5. "Computer Privacy," p. 122, Hearings of the Subcommittee on Administrative Practice and Procedure of the Committee of the Judiciary, U.S. Senate. March 14 and 15, 1967 (Washington, D.C.: U.S. Government Printing Office, 1967). (Subcommittee Chairman: Edward V. Long, Missouri.)

Chapter 22: Wired World

1. NASA's ATS-6 satellite, designed to demonstrate the uses of satellites in health care, education, and television.
2. Details of television imports and exports are from a UNESCO study, *Television Traffic–A One-Way Street?* (Paris: UNESCO Press, 1975).
3. Resolutions from the UNESCO conference in Sri Lanka in August 1976.
4. Statistic from Freedom House, a New York based organization that monitors civil rights worldwide.
5. Arthur C. Clarke, in 1945 article in *Wireless World.*
6. A Hughes 333 satellite with twelve 36-megahertz transponders manufactured by Hughes Aircraft.

INDEX